Smart Coatings

ACS SYMPOSIUM SERIES **957**

Smart Coatings

Theodore Provder, Editor
Eastern Michigan University

Jamil Baghdachi, Editor
Eastern Michigan University

American Chemical Society, Washington, DC

Library of Congress Cataloging-in-Publication Data

Smart coatings / Theodore Provder, editor, Jamil Baghdachi, editor.

    p. cm.—(ACS symposium series ; 957)

    Includes bibliographical references and index.

    ISBN 13: 978–0–8412–7429–7 (alk. paper)

    1. Protective coatings. 2. Smart materials.

    I. Provder, Theodore, 1939- II. Baghdachi, Jamil.

TA418.76.S63   2007
667′.9—dc22

                                    2006051208

Distributed by Oxford University Press

PRINTED IN THE UNITED STATES OF AMERICA

Foreword

The ACS Symposium Series was first published in 1974 to provide a mechanism for publishing symposia quickly in book form. The purpose of the series is to publish timely, comprehensive books developed from ACS sponsored symposia based on current scientific research. Occasionally, books are developed from symposia sponsored by other organizations when the topic is of keen interest to the chemistry audience.

Before agreeing to publish a book, the proposed table of contents is reviewed for appropriate and comprehensive coverage and for interest to the audience. Some papers may be excluded to better focus the book; others may be added to provide comprehensiveness. When appropriate, overview or introductory chapters are added. Drafts of chapters are peer-reviewed prior to final acceptance or rejection, and manuscripts are prepared in camera-ready format.

As a rule, only original research papers and original review papers are included in the volumes. Verbatim reproductions of previously published papers are not accepted.

ACS Books Department

Contents

Novel Coatings

Indexes

Preface

During the past 25 years, coatings technologies have been influenced by the need to lower volatile organic contents (VOC) in order to comply with stricter environmental regulations as well as to reduce the use of costly petroleum-based solvents. During this time, the use of waterborne coatings in the architectural, industrial maintenance, and original equipment manufacturing (OEM) sectors has continued to grow replacing solvent-based coatings while meeting the ever-decreasing VOC targets. In addition to waterborne coatings, other alternative technologies in the industrial and OEM sectors include powder coatings, uv-curable coatings, and high-solids coatings have had significant growth. Traditionally these coatings had the primary functions of protecting and decorating substrates. More recently, growth has occurred in research and development and commercial product generation of coatings, which have novel functions and sense and interact with their environment in addition to having the traditional protection and decoration functions. These coatings are often referred to as *smart coatings*. These types of coatings generally provide significant added value.

Smart coatings can be achieved in many ways such as by addition of additives and strategically designing polymer structures and coatings morphologies. Smart coatings categories and subtopics represented in the symposium on which this book is based are as follows:

Stimuli-responsive coatings
- coatings functioning as sensors
- thermally triggered coatings
- corrosion, degradation, and defect sensing coatings
- color shifting coatings
- light sensing coatings

Nanotechnology-based coatings
- self-assembling polymers and coatings
- photonics
- super insulating coatings
- molecular electronics
- molecular brushes
- photo-switchable materials

Novel materials and coatings
- self-repair and healing coatings
- super hydrophobic coatings
- self-lubricating coatings
- hybrid inorganic–organic coatings
- conductive materials
- self-stratifying coatings
- novel polymers and processes

Biologically active materials and coatings

- antimicrobial materials
- biocidal polymers and coatings
- photo-catalytic and bio-catalytic coatings
- antifouling and/or fouling release coatings

This book starts off with an overview of the topic of smart coatings. The first major section deals with bioactive coatings that are antibacterial as well as antifouling and fouling release coatings. The second section deals with stimuli-responsive coatings involving polymer brushes, films that respond to pH, and pressure-sensitive paints. The last section deals with novel coatings including removable conformal coatings, electro-active coatings for corrosion protection, and oxidative resistive coatings at high temperature.

We expect that this book will encourage scientific and technological investigators to expand knowledge and technology in this field as well as to apply the knowledge to commercially relevant coatings systems. We thank the authors for their effective oral and written communications and the reviewers for their helpful critiques and constructive comments

Theodore Provder, Director
Jamil Baghdachi, Professor
Coatings Research Institute
Eastern Michigan University
430 West Forest Avenue
Ypsilanti, MI 48197

Smart Coatings

Overview

Chapter 1

Twenty-First Century Materials: Coatings That Interact with Their Environment

Robert F. Brady, Jr.

706 Hope Lane, Gaithersburg, MD 20878

"Smart Coatings" represent the state-of-the-art of coatings technology. This technology has evolved through three principal stages. In the first, an ingredient with valuable properties is added to a coatings formulation, and all the unique properties of the coating are attributable to this additive. In the second, a specialized resin, pigment, or other material imparts to a coating properties that cannot be realized in any other way. "Smart Coatings," which represent the third and current stage of development, sense their environment and react to it. This paper will use this model of evolution to develop a framework for discussion at this meeting. Principles and examples of coatings at each stage of development will be given, and the way forward to unique coatings will be explored.

We've come a long way since coatings were first used some 15-25,000 years ago by Cro-Magnon people to paint figures of animals on the walls of caves in Lascaux, France. Protein binders from animal or vegetable sources were mixed with native clays and used to record the animals upon which much of life depended. As coatings became more sophisticated, they evolved from purely decorative and aesthetic applications and became essential materials for the protection and preservation of surfaces. But even as recently as 60 years ago, house painters purchased powdered white lead (lead carbonate), vegetable oil, and a drier (catalyst) of some sort, mixed them on the job site, and brushed on a protective coating that would be expected to give 3 years of service in moderate weather.

Advances in the chemical industry after World War II produced a wide variety of synthetic resins, pigments, solvents, and additives that were rapidly used to improve protective and decorative coatings. New substrate materials were also commercialized, and coatings were formulated to protect all of them. Mechanisms of protection were identified and improved. *Inhibitive coatings* such as the red lead alkyds protect steel by slowly releasing ions which passivate the metal surface. *Barrier coatings* such as epoxies and urethanes interpose a barrier between the surface and corroding species such as oxygen and water, isolating the surface from attack. *Conversion coatings* are made by phosphating or chromating a metal surface in a treatment bath. A few atomic layers of metal on the surface of the substrate are thereby converted to a hard, durable corrosion-resistant layer. *Sacrificial coatings* such as the organic or inorganic zinc coatings maintain electrical contact between a metallic substrate and the zinc metal, so that a demand for electrons will oxidize the zinc in preference to the metal of the substrate. But, basically, these coatings just sat on the surface. They were like offensive linemen on a football team: "Go ahead and give me your best shot, but you're not getting through to my quarterback!"

Now at the turn of the millennium, a coating that just "sits there" is no longer the state-of-the-art. Coatings that sense their environment and make an appropriate response are needed for forward-looking applications in medicine, aerospace, environmental protection, battlefield awareness, and personal safety, among others. So how are we to create such coatings? What principles determine their performance? How can we design a coating that perceives its environment with sufficient sensitivity, produces an acceptable response, and continues to do so during thousands, if not millions of cycles during a lifetime that must be measured in years? All the while, none of the usual restrictions are waived: the coating must be safe for workers and the environment, and – in our dreams, at least – cost no more than the material it replaces.

The name usually used to describe this new class is "Smart Coatings," and that is also the name chosen for this Symposium. "Smart" can mean many things: intelligence or mental capability, clever or witty, brash or vigorous, and trendy or fashionable. All of these features can be seen in the coatings that will be described during this Symposium.

We might consider that Smart Coatings have evolved in three stages. In the first, something that modifies coating properties all by itself is simply added to the formulation. An example of this approach would be a silicone resin, which migrates to the coating–air interface and influences the adhesion, cleanability, and/or gloss of the coating. This approach is applicable to coatings containing just about every type of resin and pigment, and the response is solely attributable to the additive itself.

A second generation of these coatings uses specialized but relatively inert ingredients to produce properties not found in more conventional coatings. Examples would include silicone or heavily fluorinated polymers, which diminish the surface energy of a coating and allow it to resist adhesion. Usually the coating has to be reformulated from scratch, for the suspension and wetting properties as well as the application and curing properties are strongly affected by the specialized ingredient.

The third stage, now a Smart Coating, is characterized by a material which truly senses and responds to its environment. The sensing element might include a pigment sensitive to light, a polymer sensitive to pH or heat, or a polymer or polymer blend which induces a patterned surface that directs wetting and spreading of liquids on its surface. The entire coating must be formulated around this unique material to ensure that it may most effectively perceive environmental stimuli and exert its characteristic response. More and more examples of this type of coating are being devised. For example, "command-destruct" coatings contain sites in the polymer backbone, which cleave in response to a stimulus, effectively destroying and removing the coating. This and other examples are given in Table I.

It is instructive to review how smart coatings technology has advanced in various applications through these three stages. Let us do this now for two dissimilar applications.

Antimicrobial and Hygenic Coatings

Health and safety issues are constantly in the public eye. People require clean food and water and are always seeking to maintain a safe and healthful living environment. Therefore facilities that supply these needs must be kept clean and sanitary at all times. Examples are waste and wastewater management plants, hospitals and clinics, industrial facilities for manufacturing and packaging pharmaceuticals and food, commercial food preparation facilities, and restaurants.

Stage one coatings are formulated by adding a nonreactive fluorinated or silicone additive to any of the epoxy, acrylic, or polyurethane coatings used for this purpose. The additive migrates to the air–coating interface and reduces the surface energy of the coating, thus impeding the attachment of bacteria and also facilitating cleaning of the surface. Alternatively, a bactericide might be simply

Table 1. Examples of "Stage Three" Smart Coatings

Coating Type	Operating Principle	Example	
		Stimulus	Response
Ablative	Outer layers of the coating surface are slowly removed by chemical (e.g., hydrolysis) or physical action	Shear of water flowing across the surface; temperature of water	"Self-cleaning:" Dirt, biofilm, etc. is removed with outer layer of polymer
Command-Destruct	Chemical bonds in the polymer backbone cleave in response to a stimulus	Heat	Depolymerizatio n and removal of the coating
Corrosion Detection	Corrosion produces OH^-; acid-base indicators change color at different pH	pH	Color reports on the site and extent of corrosion
Hygenic	Capsules contain bactericide; capsule walls are destroyed by products (e.g., acids, amines) of bacterial action	Bacteria	Release of toxin kills bacteria
Impact Sensitive	Capsules of different wall strengths contain dyes of different colors	Impact	Color reports on the site and strength of impact
Low Solar Absorption	Color pigments are transparent to UV light and allow it to be scattered by TiO_2	Sunshine	Failure to absorb UV, convert it to heat, and radiate the heat; cooler substrate
Piezoelectric	Lead–Zirconium–Titanate pigment generates an electrical current when stressed	Vibration	Proportional current; lifetime current reflects sum of stress
Pressure Sensitive	Paint contains a fluorescent pigment which is quenched by oxygen	Pressure (increased oxygen availability)	Fluorescence decreases as pressure increases
Self-Healing	Capsules containing uncured resin are broken	Impact	Replacement of damaged resin
Temperature Sensitive	Paint contains a fluorescent pigment which is quenched by heat	Temperature	Fluorescence decreases as heat increases

stirred into the coating. These "free association" coatings usually release excess active agent (*e.g.,* silicone additive or bactericide) early in their life, and fail when the amount released is too low to be effective.

Stage two coatings are based on a silicone resin or heavily fluorinated resin. These also impart low surface energy to the coating and, as the surface of the coating wears, a fresh surface of the same material is exposed and the desirable low surface energy is never lost.

A stage three smart coating contains a toxin encapsulated within various shell materials. The toxic agent is released when the shell of the capsule is destroyed. Capsules in such a coating would be made with shells sensitive to, for instance, enzymes secreted by bacteria or decomposition products (e.g., acids, amines, or ammonia) produced by bacterial action, and the rate of destruction is controlled by the chemical composition or the thickness of the shell. Capsules containing different toxins can be combined in a coating to give it broad-spectrum antimicrobial activity. The coating releases a toxin only in the presence of bacteria, preserving the active ingredient and extending its service life.

Another example of a stage three antimicrobial coating contains zeolites, an inorganic silicate which contains pores sufficiently large to hold metal ions. Silver ions can be absorbed into zeolites and then released at a controlled rate. Coatings containing silver-filled zeolites may be suitable for preventing contagion of cooking and eating surfaces and utensils, although once the silver is exhausted it may be impossible to reactivate the coating.

Antifouling Coatings

Antifouling coatings prevent the growth of marine life on the underwater surfaces of ships. This growth, or fouling, decreases the speed, range and maneuverability of a ship and raises fuel consumption by as much as 30%. Ultimately the ship must be removed from the water and mechanically cleaned to remove the fouling, and the time the ship spends out of service is also quite costly.

Early antifouling coatings were made by adding creosote or copper metal to coatings designed for other purposes. Toxic coatings representative of the second stage contained salts of arsenic, beryllium, cobalt, copper or mercury, organic biocides, or other toxins. Nontoxic second stage coatings have been created from fluorinated or silicone resins, which impart their low surface energy to the coating.

"Smart" antifouling coatings are based on an acrylic resin containing hydrolyzable esters in the side chains. The esters are transformed to alcohols at a rate, which depends on the temperature of the water and the shear imposed when water flows across the surface. The resulting polymer is water soluble and slowly washes away, taking with it biofilms and juvenile fouling organisms. These coatings usually contain toxins as well; a fresh layer of toxin is exposed as the older layers of coating dissolve away. These "ablative" antifouling coatings have been used on ships' hulls for about ten years.

Other Coatings that Interact with Their Environment

The stage three antimicrobial and antifouling coatings are good examples of smart coatings. They sense their environment and take no action until necessary. The antimicrobial coating senses bacterial degradation products and releases toxins to destroy the pathogen that produced them. The antifouling coating senses water temperature (which correlates with biological growth) and reacts with seawater, producing a water-soluble coating. What are other examples of coatings that truly sense and respond to their environment?

Lead–zirconium–titanate is a piezoelectric pigment that generates an electrical current when stressed. A coating pigmented with this material is used as a strain gauge on a pedestrian bridge. The paint is applied 50 microns (2 mils) thick directly to steel and coated with a thin conductive film. The electrodes (substrate and top film) are connected to off-site monitoring equipment, which measures the current generated by vibration and impact loads between 1Hz and 1kHz. The total current produced is a measure of the total stress experienced during the life of the structure; a sudden and sharp rise in the current may indicate a weakened structure that is less able to resist stress. The principal challenge was formulating the paint with a pigment loading high enough for the paint to function as a piezoelectric yet low enough for it to be readily applied by spray.

Self-healing coatings contain reservoirs of uncured resin within capsules. Corrosion, crack growth, mechanical damage, or chemical attack ruptures the capsules, releasing their ingredients where and when needed. The ingredients then react and become part of the film. For example, epoxies and amines (from separate capsules) react with each other; or liberated alkyd resins react with ambient air.

Medical implants are coated with a polymer, which resists adhesion to proteins, polysaccharides, fats, and cells. The polymer contains long side-chains of polyethylene oxide (PEO). The density and length of the PEO chains are tailored to function in the environment in which the implant will be used. The goal is to produce elongated, hydrated PEO chains that, in the dynamic environment of the human body, sweep across the surface of the implant and prevent settlement. The density of PEO chains on the surface is chosen so that the chains do not interfere with the motions of other chains.

Pressure-sensitive paints contain a pigment that fluoresces under ultraviolet light. Oxygen quenches this fluorescence. As air pressure on the coating increases, more oxygen is available to interfere with the output of visible light, and the intensity of fluorescence can be correlated with the pressure on the surface. This coating is used in the study of models of commercial and military aircraft in wind tunnels.

A coating that detects and reports on corrosion can be made by dispersing a number of pH indicators in a clear coating. Corrosion of metals produces hydroxide ions and raises the pH only at the site of corrosion. An acid-base indicator reacts with alkali and changes color within a specific pH range. The coating contains a selection of indicators that produce different colors and undergo their color change in different pH ranges; thus, the hue and location of color signal the extent of corrosion. Coatings can be tailored to specific substrates and to change color at a predetermined extent of corrosion.

The Future of Smart Coatings

Materials potentially useful in smart coatings are being produced in ever increasing variety and quantity. Some materials on the horizon of coatings technology are:

- Microelectromechanical devices (MEMs) are micromachined assemblies embedded below the surface of a silicon wafer. Pumps, motors, switches and other machines with dimensions of few millimeters are made by conventional photoresist technologies.
- Radio frequency identification devices (RFIDs) are millimeter-size silicon chips bearing a circuit, which oscillates in response to an external signal and responds at a chosen frequency. The signal is used to open a door, count inventory, or track parcels.
- Polymers and polymer blends form patterned surfaces. The wetting and spreading of liquids can be quite dissimilar on different polymers, and a surface patterned with polymer channels (think of a printed circuit board) can be used to direct the aggregation and flow of liquids on the surface of the coating.
- Specialized lipids self-assemble to form tubules. These tubules are plated with copper, and the lipids are removed and reused. In this way copper tubules approximately 500 nanometers in diameter and 5 micrometers long are formed. Various substances can be placed within these tubules and then slowly released from the tubules.
- Carbon nanotubes may be used as if they were nanofibers to strengthen a coating. In sufficiently high loadings, the nanotubes touch and the coating becomes conducting.

- Self-assembling nanocapsules are kept together by hydrogen bonds. Capsules in which other molecules are encapsulated may be destroyed by such strong hydrogen-bonding agents as amines and alcohols, and the internal molecule is released to exert its intended effect.
- Nanomaterials offer different chemical and physical properties than bulk materials. Although the distinction is not always precise, old nanotechnology usually describes nanoscale materials such as carbon black, fumed silica, and titanium dioxide produced for coatings for decades but getting the "nano" designer label only recently. New nanotechnology typically encompasses novel material structures, such as carbon nanotubes and quantum dots, with unprecedented properties.
- Sturdy enzymes able to resist denaturing and to function in a variety of solvents are being produced. New antimicrobial and hygenic coatings might be made by deploying these enzymes on the surface of a coating where they may breakdown bacteria or waste products.
- Polyaniline and other polymers that may be repeatedly oxidized and reduced may find a use in direct-to-metal applications such as sacrificial primers. Reducing the polymer so that it may be oxidized again may prove to be a challenge.

The way forward to produce organic surface coatings containing these radical new materials and devices is anything but certain. Some of these things must be on the surface of the cured coating, and some must be at the coating–substrate interface to ensure that they most effectively receive environmental inputs and exert their characteristic response. The coatings must be manufactured without destroying its unique properties, and must drive the device to its optimal position within the film before curing.

It's a big job, but we can dream, can't we? Coatings chemists have not yet failed any challenge, and we may look to the future with anticipation of the marvelous coatings yet to be invented.

Further Reading

1. "Antifouling Coatings Without Organotin." R. F. Brady, Jr., *Journal of Protective Coatings and Linings* **20** (1), 33-37 (2003).
2. "Antimicrobial Coatings for Food, Pharmaceutical and Hospital Facilities." R. F. Brady, Jr., *Journal of Protective Coatings and Linings* **19** (10), 59-63 (2002).
3. "Using Coatings to Manage Heat." R. F. Brady, Jr., *Journal of Protective Coatings and Linings* **19** (7), 49-50 (2002).
4. "New Developments in Fire-Resistant Coatings." R. F. Brady, Jr., *Journal of Protective Coatings and Linings* **19** (5), 54-57 (2002).

5. "Coatings that Prevent or Detect Destruction of the Substrate." R. F. Brady, Jr., *Journal of Protective Coatings and Linings* **19** (3), 65-67 (2002).
6. "Coatings that Report on Their Environment." R. F. Brady, Jr., *Journal of Protective Coatings and Linings* **19** (1), 30-31 (2002).
7. "Clean Hulls Without Poisons: Devising and Testing Nontoxic Marine Coatings." R. F. Brady, Jr., *Journal of Coatings Technology* **72** (900), 44-56 (2000).
8. "Method of Controlled Release and Controlled Release Microstructures." R. R. Price, J. M. Schnur, P. E. Schoen, M. Testoff; J. H. Georger, Jr., A. Rudolph, and R. F. Brady, Jr., U. S. Patent 6,280,759, August 28, 2001.
9. "Controlled Release Microstructures." R. R. Price, J. M. Schnur, P. E. Schoen, M. Testoff, J. H. Georger, Jr., A. Rudolph, and R. F. Brady, Jr., U. S. Patent 5,492,696, February 20, 1996.
10. M. Freemantle, "Controlling Guests in Nanocapsules." *Chemical and Engineering News* **83** (1), 30-32 (2005).

Bioactive Coatings

Chapter 2

Smart Coatings

Core-Shell Particles Containing Poly(*n*-Butyl Acrylate) Cores and Chitosan Shells as a Novel Durable Antibacterial Finish

Weijun Ye[1], Man Fai Leung[2], John Xin[1], Tsz Leung Kwong[1], Daniel Kam Len Lee[2], and Pei Li[2,*]

[1]Institute of Textile and Clothing and [2]Department of Applied Biology and Chemical Technology, The Hong Kong Polytechnic University, Hung Hum, Kowloon, Hong Kong, People's Republic of China

A novel antibacterial coating for cotton fabrics has been developed using core-shell particles that consist of poly(*n*-butyl acrylate) (PBA) cores and chitosan shells. The spherical particles were prepared via a surfactant-free emulsion copolymerization of *n*-butyl acrylate in an aqueous chitosan solution. The particles had a narrow particle size distribution with average diameter of approximately 300 nm, and displayed highly positive surface charges. Transmission electron microscopic (TEM) images clearly showed well-defined core-shell morphology with PBA cores and chitosan shells. Cotton fabric was coated with PBA-Chitosan particles through a conventional pad-dry-cure process without using any chemical binders. Fabric antibacterial efficiency was evaluated quantitatively against *Staphylococcus aureus* by using the shake-flask method. The results showed that particle-coated fabric had an excellent antibacterial property with bacterial reduction more than 99 %. The durability of the antibacterial property was maintained above 90 % bacterial reduction even after 50 times of home laundering. Effects of particles coating on fabric hand, air permeability, break tensile strength and elongation, as well as fabric surface morphology, were investigated.

Introduction

Natural textiles such as those made from cellulose and protein fibers are considered to be vulnerable to microbe attack because of their hydrophilic porous structure and moisture transport characteristics. Thus, the use of antibacterial agents to prevent or retard the growth of bacteria becomes a standard finishing for textile goods. On the other hand, there is an increasing public concern over the possible effects of antibacterial finishing on environmental and biological systems since many antibacterial agents are toxic chemicals, and exhibit a lack of efficiency and durability (1,2,3). Hence, an ideal textile antibacterial finishing should not only kill undesirable microorganisms and stop the spread of diseases, but also fulfill three other basic requirements: 1) Safety. The product should not be excessively toxic to human and the environment, and should not cause skin allergy and irritation. 2) Compatibility. The product must not negatively impact to textile properties or appearance and must be compatible with common textile processing. 3) Durability. The product should be able to endure laundering, drying and leaching. Thus, there is an increasing research effort that is directed to the development of natural antibacterial coatings that are safe, durable and environmentally friendly.

Chitosan, a β-(1,4)-linked polysaccharide of D-glucosamine, is a deacetylated form of chitin, the second most abundant natural polymer in the world. They are obtainable from the shells of crabs, shrimps and other crustaceans. Chitosan is a non-toxic, biodegradable and biocompatible natural polymer, and has long been used as a biopolymer and natural material in diverse applications (4,5,6). Because of its polycationic nature, chitosan possesses good antibacterial properties against various bacteria and fungi through ionic interactions at a cell surface which eventually kills the cell (7,8). Past studies have indicated that its antimicrobial activity was influenced by molecular weight (9), degree of deacetylation (10), temperature, pH and cations in solution (11).

Since chitosan is one of the safest and most effective antibacterial agents, it has been widely used for cotton and other textile antibacterial finishes (12,13,14,15). However, direct coating of the chitosan onto textile articles suffers from drawbacks such as its insolubility in most solvents except acidic solutions; high viscosity of chitosan in aqueous solution causing many handling problems; and poor fabric hand after coating due to rigid nature of the chitosan. To address these problems with the use of chitosan, various methods have been developed to chemically modify the chitosan (16). For instance, functional groups such as quaternary ammonium salts, were introduced to the chitosan backbone in order to improve its solubility in water and antibacterial activity (17,18,19). Completely deacetylated chitosan in sodium nitrite was coated on cotton and gave a better laundry durability (20). Chitosan was grafted onto poly(ethylene terephthalate) fibers or polyhydroxyalkanaote membranes through plasma glow discharge or ozone treatment to produce antimicrobial materials (*14,21*). In this paper, we present a new approach to tackle aforementioned problems associated with the use of chitosan. Core-shell particles that consist of soft polymeric cores and chitosan shells have been designed and synthesized. Their potential application in antibacterial coating of textiles was evaluated.

Experimental

Materials

Chitosan (Aldrich) was purified by dissolving it in a 0.6% acetic acid solution at 60 ^{0}C, followed by filtration and precipitation in 10% NaOH solution under stirring at room temperature. The chitosan was then filtered and washed with distilled water until it reached a neutral pH, and dried in a vacuum oven at 60 ^{0}C. Molecular-weight measurement, based on solution viscosity, suggested that the M_v of chitosan was approximately 80,000 g/mol. Its degree of deacetylation, as estimated by ^{1}H NMR spectroscopy (*22*), was 74%. *n*-Butyl acrylate (BA, Aldrich) was purified by vacuum distillation. *Tert*-butyl hydroperoxide (*t*-BuOOH, 70% aqueous solution) was obtained from Aldrich and diluted with water to 20 mM as a stock solution. Acetic acid (Riedel-de Haën, Germany) was used as received. Freshly deionized and distilled water was used as the dispersion medium. Scoured and bleached plain woven cotton fabrics (100%) were rinsed with nonionic detergent before finishing.

Synthesis of PBA-Chitosan Particles

The core-shell particles were prepared via a surfactant-free emulsion copolymerization according to our previously described method with minor modifications (*23*). A 250 mL round-bottomed, three-necked flask equipped with a condenser, a magnetic stirrer and a nitrogen inlet was immersed in an oil bath. In a typical run, 100 mL of 0.6% acetic acid solution was added to the flask, followed by the addition of purified chitosan (0.5 g). The mixture was stirred at 60 °C to completely dissolve the chitosan. After purging nitrogen for 30 min at 80 °C, *n*-butyl acrylate (2 g) was added dropwise, followed by a quick addition of 1 mL of *t*-BuOOH solution (2.0 x 10^{-2} M). The polymerization was conducted for 5 h at 80 °C under nitrogen atmosphere. Upon completion, the white latex dispersion was cooled to room temperature and stored for finishing procedure. The monomer conversion was determined gravimetrically.

Measurement and Characterization

Infrared spectra of polymers were recorded on a FT-IR spectrophotometer (Nicolet 750) using KBr disks. Infrared spectra of coated cotton fabrics were recorded on a reflective FT-IR spectrophotometer (Perkin Elmer). Particle size and size distribution were measured on a Particle Size Analyzer (Coulter LS-230). The *zeta*-potentials of particles were measured with a Zetasizer 3000HS (Malvern) with a 1 mM NaCl aqueous solution as the suspension fluid. The core-shell nanostructures of the particles were observed using a scanning transmission electron microscope (STEM, *FEI* Tachai 12) at an accelerating voltage of 120 kV after treating the particles with 2% phosphotungstic acid for an appropriate period. The morphologies of cotton fabrics before and after coating were examined by scanning electron microscopy (SEM, Stereoscan 440). Cotton fabrics were cut into small pieces and fixed on the standard SEM sample holders with a double-coated carbon conductive tab. All samples were sputter coated with a thin layer of gold in a vacuum.

Mechanical properties of cotton samples before and after the treatment were measured with KES-FB instruments under a standard temperature and a moisture condition (20 °C, 65 % RH). Air permeability was performed with an automatic air-permeability tester (KES-FB-AP1, Kato Tech Co.) by measuring air-resistance of a constant air flow through a fabric specimen from and into the atmosphere. Bending rigidity and hystersis of bending moment were determined with a Pure Bending Tester (KES-FB2, Kato Tech Co.) in order to evaluate properties of fabric hand and stiffness. Tested samples were cut into 20 x 20 cm squares. Tensile strength was determined with an Instron Tensile Tester-4411. Breaking strength and elongation of test strips (2.5 cm in width) along both warp and weft directions were recorded.

Antibacterial Coating

The antibacterial finish was applied through the conventional pad-dry-cure method. Each fabric sample ($\sim 20 \times 40$ cm) was washed with nonionic detergent, and immersed in the latex dispersion for 3 to 5 minutes. The sample was then put through a laboratory pad machine (Rapid Vertical Padder, Taiwan) under a nip pressure of 1 kg/cm^2 for a complete wet pick-up. The dip-pad procedure was repeated twice, and the padded samples were dried in an oven at 100 °C for 5 minutes, and cured at 150 °C for 4 minutes. After rinsing, the treated samples were dried again.

Antibacterial Activity

The antibacterial activity was evaluated quantitatively by using a shake-flask method developed by Dow Corning Corporation (*24*). This method is specially designed for specimens treated with non-releasing antibacterial agents under dynamic contact conditions. The test determines the reduction in the number of bacterial cells after placing the sample in a shaking flask for 1 h. *S. aureus* (ATCC 6538), a gram positive bacterium commonly found on the human body, was chosen as the tested bacterium. The following is a typical procedure: 1 ± 0.1 g of sample fabric, cut into small pieces (approximately 0.5×0.5 cm), was placed into a flask containing 50 mL of 0.5 mM monopotassium phosphate (PBS) culture solution with a cell concentration of $1.0 - 1.5 \times 10^4$/mL. The flask was then shaken for 1 h at 250 rpm on a rotary shaker at 37 °C. Before and after shaking, 1 mL of the test solution was extracted, diluted and spred on an agar plate. After incubation for 24 h at 37 °C, the number of colonies formed on the agar plate was counted and the number of live bacterial cells in the flask was calculated. Antibacterial efficacy was determined based on duplicated test results. Percentage bacterial reduction was calculated according to the equation of $R = (B-A)/B \times 100\%$, where R is the percentage bacterial reduction, B and A are the numbers of live bacterial cells in the flask before and after shaking, respectively.

Laundering Durability

The antimicrobial durability of the particle-coated cotton fabrics was investigated with an accelerated wash fastness test according to AATCC Test Method 61-1996. An AATCC standard washing machine (Atlas Launder-Ometer) and detergent (AATCC Standard Detergent WOB) were used. Samples were cut into 5 x 15 cm swatches and put into a stainless steel container with 150 mL of 0.15 w/v % WOB detergent solution and 50 steel balls (0.25 in. in diameter) at 49 °C for various washing times to simulate 5, 20 and 50 wash cycles of home/commercial launderings.

Results and Discussion

Preparation and Chacterization of PBA-Chitosan Core-Shell Particles

A new method to prepare well-defined, core-shell particles that consist of hydrophobic cores and hydrophilic shells has been developed via a surfactant-free emulsion copolymerization (*23,25*). This versatile approach allows us to design and synthesize a variety of core-shell particles for specific applications (*26, 27*). In this research, we have designed a novel core-shell particle consisting of a poly(*n*-butyl acrylate) core and chitosan shell as a potential antibacterial coating agent (*28*). Since PBA is a soft polymer, it is anticipated that the coatings with PBA-Chitosan core-shell particles on textiles will possess notable features such as antibacterial property, good fabric hand and durability.

Poly(*n*-butyl acrylate)-Chitosan core-shell particles were prepared using our previously described method (*23*). Chitosan-*graft*-PBA copolymers were generated during the polymerization and formed as shell layer of the particles. Dynamic light scattering measurement indicated that these particles had a number-average particle diameter of 320 nm with narrow size distribution (particle size distribution, D_v/D_n = 1.16, where D_v and D_n are the volume- and number-average particle diameters, respectively) (Figure 1). Surface charges of the PBA-Chitosan particles were studied through *zeta*–potential measurements as a function of pH in 1 mM NaCl solution at 25 ^{0}C as shown in Figure 2. The positive potential decreased as the pH increased, indicating that cationic chitosan molecules were located on the particle surfaces. When the pH of the dispersion was higher than 8.5, the particles became unstable due to low surface charge density. Figure 3 illustrates transmission electron microscopy (TEM) micrographs of PBA-Chitosan particles. They are spherical and have narrow size distribution (Figure 3a). A well-defined core-shell morphology where the poly(*n*-butyl acrylate) cores are coated with chitosan shells is clearly revealed. Figure 3(b) displays the deformation of the particles when in contact with each other, indicating that the PBA-Chitosan particles are very soft.

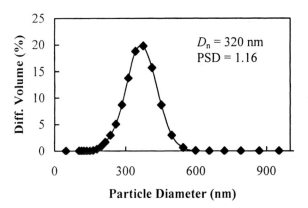

Figure 1. Dynamic light-scattering measurement of average size of PBA-Chitosan particles and its size distribution (PSD = D_v/D_n). (Reproduced with permission from reference 29. Copyright 2005 Elsevier.)

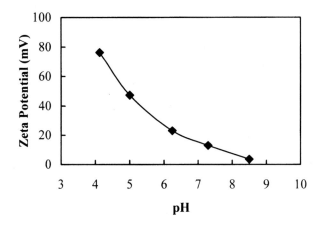

Figure 2. Surface charges of PBA-Chitosan particles as a function of pH in a 1 mM NaCl solution at room temperature (Reproduced with permission from reference 29. Copyright 2005 Elsevier.)

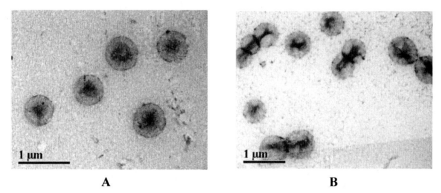

A   B

Figure 3. TEM micrographs of PBA-Chitosan particles negatively stained with 2% PTA solution for 30 mins. (A) Well-defined core-shell morphology with PBA cores and chitosan shells; (B) Soft PBA-Chitosan particles which deform easily when in contact with each other (Reproduced with permission from reference 29. Copyright 2005 Elsevier.)

Antibacterial Coating

The PBA-Chitosan particle dispersion synthesized above was applied to a pure cotton fabric by a conventional pad-dry-cure method as described in the experimental scetion. The successful coating of particles on the fabric was confirmed with Reflective FT-IR spectra. A small carbonyl absorption peak at 1730 cm^{-1} in the treated cotton sample indicated the presence of ester groups of PBA. The morphology of the fabric surfaces before and after coating with the particles was examined with SEM (Figure 4). There was little difference in their surface appearance, and no individual particles were observed on fabric surface. The smooth coating of the PBA-Chitosan particles is attributed to the good film-forming property of the PBA polymer.

Antibacterial Activity

It is well recognized that chitosan has good antimicrobial activity, especially against the growth of *Staphylococcus aureus* (*S. aureus*) *(13)*. Thus, bacterial reductions of particle-coated and chitosan-treated cotton fabrics were evaluated using a Shake-Flask Method as described in the experimental section. It was found that untreated fabric gave a negligible antibacterial activity (less than 5 %), while all finished cotton samples showed over 99% bacterial reduction. Comparable bacterial reduction (> 99 %) was also obtained for samples treated with 0.5 wt % chitosan solutions. Thus the graft copolymerization of *n*-butyl acrylate from the chitosan did not affect chitosan's antimicrobial property.

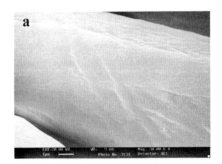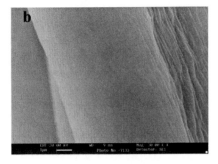

Figure 4. SEM micrographs of surface morphologies of: (a) untreated cotton fabric; (b) particle-treated cotton fabric (Reproduced with permission from reference 29. Copyright 2005 Elsevier.)

Mechanical Properties

To evaluate the effect of latex coatings on the mechanical properties of the cotton fabric, fabric hand, air permeability, and tensile strength of the cotton before and after coating with particles and chitosan were examined (30). Bending rigidity (B) and hysteresis of bending moment (2HB) are two parameters related to fabric stiffness and difficulty of fabric deformation under bending. Results in Table 1 show that coating with particles or chitosan increases both B and 2HB values in machine (warp) and cross (weft) directions. Although both coated cotton fabrics became stiffer after treatment, the particle coated-cotton has much lower B and 2HB values, suggesting that it has a better fabric hand than the chitosan-treated one.

Air resistances (R) of the cotton fabric before and after coating were measured. The R value represents an air pressure required to keep a constant airflow penetrating the fabric surface. A smaller resistance value means a better air permeability. It was found that the R value of the cotton fabric dropped from 1.47 KPa·S/m to 0.68 KPa·S/m after treatment with PBA-Chitosan particles. The results suggest the smoothing of fabric surface with the PBA-Chitosan particles. Lee et al. observed a similar improvement on fabric air permeability after coating chitosan on cellulose, silk, and wool (31).

To evaluate the effect of antibacterial coating on fabric resistance to stretching or pulling force, fabric strips were tested for their breaking and elongation properties. Results in Table 2 show that fabric tensile strengths of both warp and weft directions slightly decrease after the coating. But all fabrics still maintain at least 77 % of their original tensile strength.

Table 1. Bending Rigidity (B) and Bending Hysteresis (2HB) of Cotton Fabric Before and After Coating

Direction	Warp		Weft	
Sample	B-MEAN [gf.cm²/cm]	2HB-MEAN [gf.cm/cm]	B-MEAN [gf.cm²/cm]	2HB-MEAN [gf.cm/cm]
Before coating	0.0353	0.0360	0.0720	0.0658
Coated with PBA-Chitosan	0.0604	0.0310	0.1756	0.0796
Coated with chitosan	0.0783	0.0508	0.2860	0.1361

Table 2. Break Tensile Strength and Elongation of Cotton Fabric Before and After Coating with PBA-Chitosan Particles

Tensile Direction	Warp		Weft	
	Break force (N)	*Break strain (%)*	*Break force (N)*	*Break strain (%)*
Before coating	377.5	17.8	219.2	15.0
After coating	364.6	20.7	168.6	12.3

Laundering Durability

One of the biggest concerns of antibacterial finish in today's textile industry is the durability. An ideal antimicrobial finish should be effective for the entire lifetime of a textile article. Generally, if a textile material can maintain at least 80 % of its inhibitory activity after twenty times of home laundering, it is considered to be a "durable antibacterial finish" (*3*). To evaluate the durability of the antibacterial finish, a washfastness test on cotton fabric was performed according to the AATCC standard Method (61-1996). The efficacy was determined by measuring the fabric bacterial reductions after 0, 5, 20, and 50 repeated wash cycles. The washfastness results are illustrated in Figure 5 as the percentage of reductions in the number of *S. aureus* cells. All washed samples maintain high antibacterial efficacy even after 50 times of repeated laundering. The washfastness test in acidic human sweat with a pH of 4.5 was proformed for 20 and 50 wash cycles. High antibacterial activity (over 90% bacterial reductions) was achieved. Such an excellent antimicrobial durability means that without an additional chemical binder, the coated PBA-Chitosan particles on fabric surface could not be washed away, even under an acidic environment. Thus the PBA-Chitosan particles adhered firmly onto the cellulose surface, most likely through both physical interactions and certain chemical bindings.

Conclusion

Novel chitosan-based core-shell particles, with chitosan as the shell and a soft polymer as the core, have been designed and synthesized as an antibacterial coating for cotton fabric. The core-shell particles were prepared via a graft copolymerization of *n*-butyl acrylate from chitosan in aqueous solution. Quantitative antibacterial tests of the coated cotton fabrics demonstrate an

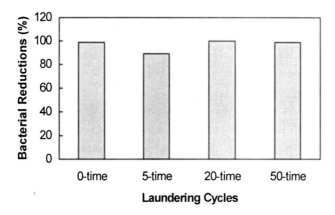

Figure 5. Antibacterial activity of treated cottons in terms of percentage bacterial reduction after different laundering cycles

excellent fabric antibacterial activity with 99 % bacterial reductions. The particle coating is also durable since over 90 % bacterial reduction can still be maintained even after 50 times of laundering cycles. The particle-treated fabric has a better air permeability with little reduction in fabric tensile strength. The fabric surface morphology is not affected by the particle coating due to the softness of the particles. At the same time, it results in a better fabric hand than the one coated with chitosan only.

Acknowledgment

We gratefully acknowledge the Hong Kong Polytechnic University (A-PE72) and Innovation Technology Fund of the Hong Kong Government for their financial support of this research.

References

1. Mao, J.; Murphy, L. *AATCC Review* **2001**, *1*, 28.
2. Bohringer, A.; Rupp, J.; Yonenaga, A. *Inter. Text. Bull.* **2000**, *46*, 12.
3. Corcoran, L. M. *Determining the Processing Parameters and Conditions to Apply Antibacterial Finishes on 100 % Cotton and 100 % Polyester Dyed Knit Fabrics*; Institute of Textile Technology, Charlottesville, Virginia, 1998.

4. Sashiwa, H., Aiba S. *Prog. Polym. Sci.* **2004**, *29*, 887.
5. Sandford, P. A. *Chitin and Chitosan*, Elsevier Science Publishers: London and NY, 1989; pp. 51-69.
6. Muzzarelli, R. A. A. *Natural Chelating Polymers*, Pergamon Press: NY, 1973.
7. Jolles, P.; Muzzarelli, R. A. A. *Chitin and Chitinases*, Birkhauser Verlag: Basel, 1999; pp. 315-333.
8. Shin, Y.; Yoo, D. I.; Min, K. *J. Appl. Poly. Sci.* **1999**, *74*, 2911.
9. Shin, Y.; Yoo, D. I.; Jang, J. *J. Appl. Poly. Sci.* **2001**, *80*, 2495.
10. Liu, X. F.; Guan, Y. L.; Yang, D. Z.; Li Z.; Yao, K. D. *J. Appl. Poly. Sci.* **2001**, *79*, 1324.
11. Tsai, G. J.; Su, W. H. *J. Food Protect.* **1999**, *62*, 239.
12. Oktem, T. *Color Technol.* **2003**, *119*, 241.
13. Lee, S.; Cho, J. S.; Cho, G. *Textile Res. J.* **1999**, *69*, 104.
14. Huh, M. W.; Kang, I. K.; Lee, D. H.; Kim, W. S.; Lee, D. H.; Park, L. S.; Min, K. E.; Seo, K. H. *J. Appl. Poly. Sci.* **2001**, *81*, 2769.
15. Xu, Y.; Chen, J. *Fangzhi Xuebao* **2001**, *23*, 13.
16. Lim, S. H.; Hudson, S. M. *J. Macromol. Sci. Part C Polym. Rew.* **2003**, *43*, 223.
17. Kim, Y. H.; Choi, H. M.; Yoon, J. H. *Textile Res. J.* **1998**, *68*, 428.
18. Kim, Y. H.; Nam, C. W.; Chio, J. W.; Jang, J. *J. Appl. Poly. Sci.* **2003**, *88*, 1567.
19. Seong, H. S.; Whang, H. S.; Ko, S. W. *J. Appl. Poly. Sci.* **2000**, *76*, 2009.
20. Seong, H. S.; Kim, J. P.; Ko, S. W. *Textile Res. J.* **1999**, *69*, 483.
21. Hu, S. G.; Jou, C. H.; Yang, M. C. *J. Appl. Poly. Sci.* **2003**, *88*, 2797.
22. Hirai, A.; Odani, H.; Nakajima, A. *Polym. Bull.* **1991**, *26*, 87.
23. Li, P.; Zhu, J.; Sunintaboon, P.; Harris, F. W. *Langmuir* **2002**, *18*, 8641.
24. *Corporate Test Method 0923*; Dow Corning, 1979, Midland, Michigan.
25. Li, P.; Zhu, J.; Sunintaboon, P.; Harris, F. W. *J. Disp. Sci. Technol.* **2003**, *24*, 607.
26. Zhu, J.; Li, P. *J. Polym. Sci. Part A: Polym. Chem.* **2003**, *41*, 3346.
27. Zhu, J.; Tang, A.; Law, L. P.; Feng, M.; Ho, K. M.; Lee, D. K. L.; Harris, F. W.; Li, P. *Bioconjugate. Chem.* **2005**, *16*, 139.
28. Ye, W.; Leung, M.F.; Xin, J., Kwong, T.L.; Lee, D. K.-L.; Li, P. *Polymer* **2005**, *46*, 10538.
29. Reprinted from *Polymer*, Vol. 46, Ye, W.; Leung, M.F.; Xin, J., Kwong, T.L.; Lee, D. K.-L.; Li, P., Novel Core-Shell Particles with Poly(*n*-butyl Acrylate) Cores and Chitosan Shells as an Antibacterial Coating for Textiles, Page 10538, **2005**, with permission from Elsevier.
30. Ye, W.; Xin, J.; Li, P.; Lee, D. K.-L.; Kwong, T.L. *J. Appl. Polym. Sci.* **2006**, In Press.
31. Lee, H. J.; Jeon, D. W. *J. Korean Fiber Soc.* **1999**, *36*, 478.

Chapter 3

Formulation and Evaluation of Organic Antibacterial Coatings

D. L. Clemans[1], S. J. Rhoades[1], J. J. Kendzorski[1], Q. Xu[2], and J. Baghdachi[2]

Departments of [1]Biology and [2]Coatings Research Institute, Eastern Michigan University, Ypsilanti, MI 48197

Many infectious diseases can be acquired through casual contact with infected individuals and environmental sources. Routine sanitation by the use of disinfecting agents is fairly successful in combating the spread of some infectious disease. However, cleaning procedures can be costly and time consuming, and do not offer continuous protection or guard against fresh contamination. Self-disinfecting surface coatings containing non-migratory antimicrobial agents can fulfill this task. The use of such surface coatings containing non-migratory antimicrobial agents, while not aimed at producing a totally sterile environment, may serve a useful purpose in areas where a permanent and more sanitary environment is necessary. This paper overviews our recent results on the formulation and testing of model antimicrobial surface coatings. The antimicrobial activities of these coatings on the test glass substrate were tested toward a selected panel of gram-positive (*Staphylococcus aureus* ATCC 6538) and gram negative (*Escherichia coli* ATCC 11229) bacteria. We have demonstrated that air dry waterborne coatings containing silver-based antimicrobials effectively kill the test strains of *E. coli* and *S. aureus* at a concentration of 10^5 CFU/ml after 12hr of incubation at RT. Even with increased bacterial loads (10^7-10^9 CFU/ml) >95% killing of the test organisms on the antibacterial coatings have been observed. The proposed possible applications of such coatings include areas such as clinical settings, food preparation areas, and food processing facilities. The challenge with such applications is to determine the ability of coatings to kill bacteria while having a subsequent layer of an organic nature.

Introduction

Many infectious diseases can be acquired through casual contact with infected individuals and environmental sources. These environments include different hospital and office surfaces, food and drug processing facilities, meeting facilities in schools; and private homes. Despite the best available sanitation and use of antimicrobial agents, many infectious diseases are acquired through contact with various surfaces (i.e., fomites). The contamination of fomites occurs through the aerosolization and deposition of various infectious agents (e.g., bacteria, viruses, fungi) on inanimate objects such as door handles, counter tops, etc. Casual contact with these fomites spread the infectious agents from one person to another.

While routine sanitation is fairly successful, it is costly and does not offer continuous protection and more importantly, no protection against fresh contamination. Self-disinfecting surface coatings containing non-migratory antimicrobial agents can fulfill this task. The use of antimicrobial surface coatings while not aimed at producing a totally sterile environment may serve a useful purpose in areas where a permanent and more sanitary environment is necessary.

Cleaning supplies, disinfecting agents and labor are costly, and the constant need to clean takes time away from other productive activities. It would be desirable to reduce these expenses by painting surfaces with antimicrobial coatings that fight fungi and bacteria upon contact. However, the coating is not a completely sterile environment and will not be effective against some strains of bacteria and viruses, so the need for cleaning will not completely disappear.

Coatings can be designed to kill bacteria through three mechanisms: 1) coatings that can resist the attachment of bacteria, 2) coatings that release biocides that will kill the bacteria, and 3) coatings that kill bacteria on contact (1-4). Coatings can also combine two or more of these mechanisms.

Coatings that Resist Adhesion

Bacteria produce biofilms that are hospitable environments where they can grow without difficulty. Coatings that prevent or resist the formation and growth of the biofilm are very promising agents to combat bacteria and fungi. To accomplish this, the surface must be hydrophobic. To date efforts to use hydrocarbon or fluorocarbon additives in hydrophobic coatings have not been successful; the oily additives in hydrophobic coatings eventually migrate from the coating, and the solid additives tend to create porous surfaces.

Hydrogels are the most important family of hydrophilic adhesion-resistant coatings. Hydrogels based on polyurethane chemistry are widely used in medicine to prevent the adhesion of cells to vascular implants and artificial joints. In practice, however, their use as surface coatings is limited by their comparatively poor resistance to abrasion and commercial cleaners.

Coatings that Release Toxins

The second family of antimicrobial coatings are those that release compounds that kill microorganisms. Many suitable bactericides and fungicides are available commercially, including antibiotics, quarternary ammonium compounds ("quats"), and other organic materials. The bactericides commonly used, as in-can stabilizers for latex coatings, such as chloramines, triazines, ethylene oxide, or phenols, have no value here. The service life of these coatings is primarily determined by how long an effective release of toxin can be sustained. Simply mixing a toxin into a coating usually produces an inefficient, costly, short-lived, and relatively ineffective coating. The agent is released too quickly when the coating is new, wasting an expensive ingredient. Soon, the amount of agent released is too low to be effective.

Extensive open literature and patent search indicate that there are scores of chemical agents that can destroy certain microorganisms. The chemical make up of such compounds are well known and range from organic compounds such as quaternary ammonium compounds, N-alkylated poly(4-vinyl-pyridine), active on both gram-positive and gram-negative bacteria (5-7) Arylamide polymers (8) calcium hydroxide to encapsulated silver ions (1-4, 9-12).

When used in coatings, each class of compounds has some advances and some limitations. Organic compounds that are not bound to the backbone resin of the coating can leach out over time greatly reducing its useful life. It has been well documented that some microorganisms can develop resistance to or alters chemical properties of the antimicrobial agent(s). The lifetime of coatings containing bactericide as an unbound organic additive therefore; may depend on the applied thickness, the frequency and intensity of cleaning and bacterial resistance.

Zeolites are inorganic silicates that contain pores large enough to hold metal ions. Silver ions can be absorbed into zeolites and then released at a controlled rate. Coatings containing silver filled zeolites may be suitable for controlling and preventing the growth of a wide variety of microorganisms. One specific agent is zirconium phosphate complex of silver ions which as found commercial application as an antimicrobial agent in textile, filtration media, building materials and adhesives.

Silver has a very long history of use as an effective, broad-spectrum antimicrobial agent (13-18). Silver was used as an antimicrobial agent at least as long ago as 1000 BC (19) in applications ranging from water treatment to medicine. Silver is used today to impart bactericidal properties to water filtration devices, as an alternative to halogen-based products in swimming pools, spas and small cooling water systems (20-23), and to control infections in hospitals (19, 21, 23). Clinical applications include Ag-sulphadiazine for treatment of burns (24), $AgNO_3$ for prevention of gonorrhea opthalmicum in neonates (23), and Ag-complexes for dental resin composites (25). The reasons for the historical use of silver as a biocide are based on the relative safety to mammals and efficacy against a broad range of microorganisms.

Over the last 100 years, many studies have been published where the antimicrobial activity of silver ions has been quantified. Recent papers suggest that the mode of action of silver ions varies depending upon the target organism, and can interact with functional groups on proteins containing sulfur, phosphorus, nitrogen, and oxygen (26). Further, silver ions have been shown to disrupt the TCA cycle, electron transport system, phosphate uptake, transport of various metabolites, and the polarization of bacterial membranes (26).

Recent studies have elucidated a broad spectrum of heavy metal resistance mechanisms in both clinical and environmental isolates (27-34). Bacterial resistance to heavy metals such as silver ions appears to be through the acquisition of various efflux pumps that pump the silver ions out of the cell (27-29). These genes, however, are different from those that mediate resistance to various antibiotics. Silver-based products have been used extensively in Japan and in Europe. No reports of the failure of these products have been published.

There is a major gap in understanding of antimicrobial activities where continuous exposure to environment exists. Another challenge is that the antimicrobial agents will need to be retained at the targeted areas for as long as possible in order to provide sustained antimicrobial activities and should not cause bacterial resistance. In addition, the antimicrobial agent(s) must be resistant for mechanical removal and remain biologically active in the presence of a variety of complex chemicals such as variety of oils, protein, aqueous and non-aqueous cleaning products.

There is no systematic study that correlates the activity of such compounds as a function of multitude of parameters, in particular in home, hospitals and in personal care applications. This investigation focuses on making simple antimicrobial formulations and testing the coating activities on selected microorganisms.

The antimicrobial activities of these coatings on the test substrates are tested toward a selected panel of gram-positive (*Staphylococcus aureus* ATCC 6538) and gram negative (*Escherichia coli* ATCC 11229) bacteria. These strains of bacteria have been used as standards for testing of disinfectant efficacy using the AOAC Use-Dilution Methods and represent common bacteria encountered in health care and food preparation facilities (35,36, 37-40).

Experimental

Antimicrobial coatings were formulated by incorporating various levels encapsulated silver containing antimicrobial agents in the appropriate resin systems. The antimicrobial coating formulations include both solvent borne and waterborne coatings. The resins used in the formulations of the antimicrobial coatings include acrylic thermoplastic solution Chempol 317-0066, CCP, acrylic emulsion Neocryl A622, Neoresins Inc., Flexbond 381, Air Products, water-reducible polyester dispersion Bayhydrol XP 7093, Bayer Material Science and urethane/acrylic copolymer dispersion, Hybridur 570, Air products. Additionally, the formulations contain various additives, pigments and solvents.

Air dry and thermosetting two-component coatings are prepared in conventional manner. Waterborne resins that contained strong sulfur and chloride anions required special treatments. Formulations reported in this investigation contained the silver-based antimicrobial agents in the range of 0.4-6.0% based on the total formulation weight. The particle sizes of the antimicrobial agents were reported by the suppliers to be in the ranges of 1-5 microns.

Panel Preparation and Physical Testing

The coatings were applied onto glass, steel and aluminum panels using drawdown rod to produce dry films in the range of 12-50 microns. Air dry and waterborne coatings were allowed to dry at room temperature for a minimum of 7 days before physical and biological testing. Thermosetting coatings were cured depending on the crosslinking chemistry at 80°C x 30 min. or 140°C x 30 min.

All coatings were tested for key physical properties before submitting for antimicrobial activity testing. The solvent resistance was tested according to ASTM D 5402-93 using 70% ethanol in water as the solvent. The tape adhesion test was done according to ASTM D 3359-92a. The film hardness was determined using pencil hardness test according to ASTM D 3363. The film thickness was tested according to ASTM D 1400. The results are shown in Table I.

Table I. Key Physical Properties of Prototype Antimicrobial Coatings.

Run	Adhesion	Film Thickness	Pencil Hardness	70% EtOH Resistance	Appearance	Comments
01	5B	45	3B	>200rubs	Clear	Latex paint
1_A	5B	50	3B	>200rubs	Hazy	
1_B	5B	42	3B	>200rubs	Clear	
02	5B	45	3B	>200rubs	Clear	Water-reducible
2_A	5B	45	3B	>200rubs	Clear brown*	
2_B	5B	50	3B	>200rubs	Clear pink*	
03	4B	40	2B	>200rubs	White	Pigmented Latex
3_A	4B	45	2B	>200rubs	Light purple*	
3_B	4B	45	2B	>200rubs	White	2% loading
3_B	3B	45	3B	>200rubs	White	6% loading
04	5B	40	2H	>200rubs	Blue	Two part
4_A	5B	45	2H	>200rubs	Blue	Polyurethane
05	5B	50	2H	>200rubs	Green	Acrylic, baked at
5_A	5B	50	2H	>200rubs	Green	140 °C

01: Control

$1_{A,B}$: Antimicrobial agents

*Changed color on standing

Evaluation of Organic Antimicrobial Coatings
Materials and Methods

Sterilization of coupons

The antibacterial coatings to be tested were supplied as a single-sided coating on standard 1" X 2" glass microscope slides. Sterilization of the coupons to rid them of any contaminating microorganisms prior to the challenge testing was achieved after exposure of the coupons to the germicidal UV lamp in a Nuaire Model NU-455-600 Class II Type A/B3 laminar flow hood (2 min. on each side at RT).

Preparation of Bacterial Broth Cultures

Esherichia coli ATCC11229 (Culti-loops, Remel Europe, Dartford, Kent, UK) and *Staphylococcus aureus* ATCC6538 (Culti-loops, Remel Europe, Dartford, Kent, UK) were selected for the challenge testing of the antimicrobial coatings. Each bacterial strain was streaked for isolated colonies on Tryptic Soy agar plates (TSA; Difco, Becton Dickinson, Sparks, MD) and incubated at 37°C for 12 to 18 hours. Broth cultures were prepared by first picking 1-2 isolated colonies from TSA plates and inoculating 5 ml Tryptic Soy Broth (TSB; Difco, Becton Dickinson, Sparks, MD) in a large test tube. This culture was grown in a shaking water bath for 8 hours at 37°C. After the eight hours, 125 ul of this culture is added to 50 ml of TSB in a side-arm Klett flask and grown for approximately 8 hours in shaking water bath at 37°C. The bacterial broths were grown to an A_{590} of 1.2 which correlated to approximately 10^9 CFU (colony forming units)/ml. 25 ml of this broth was transferred to a sterile 30 ml centrifuge tube and centrifuged at 10,000 rpm for 10 minutes at 4°C (Sorval SS34 rotor) to pellet the bacteria. The bacterial pellet was washed one time and re-suspended in 25 ml of Phosphate Buffer Solution (PBS; BBL, FTA Hemagglutination Buffer, Becton Dickinson, Sparks, MD). The suspension was diluted with PBS to achieve the target range of 10^5-10^6 CFU/ml and used to inoculate the coupons containing antibacterial coatings.

Application of Bacteria to Coating

All inoculations were performed in a laminar flow hood. Sterilized coupons were placed into a sterile petri dish. 0.5 ml of prepared bacterial suspensions was pipetted directly onto the center of each coupon which formed a drop approximately 1.5 cm in diameter. A total of 36 coupons were tested in each individual run of the experiment; Table II details the experimental design for each coating tested.

Table II. Design of a Typical Antibacterial Coating Challenge Test.

Inoculum	0 hour[1, 2]	12 hour[1, 2]	24 hour[1, 2]
E. coli ATCC11229	2 w/AMA, 2 w/o AMA	2 w/AMA, 2 w/o AMA	2 w/AMA, 2 w/o AMA
S. aureus ATCC6538	2 w/AMA, 2 w/o AMA	2 w/AMA, 2 w/o AMA	2 w/AMA, 2 w/o AMA
PBS only	2 w/AMA, 2 w/o AMA	2 w/AMA, 2 w/o AMA	2 w/AMA, 2 w/o AMA

[1] w/AMA = coating with the antimicrobial agent
[2] w/o AMA = coating without antimicrobial agent

Once the coupons were inoculated, those assigned to incubate for 12 or 24 hours were placed in a humidified chamber. This was achieved by placing the petri dishes containing the coupons on test tubes that had been taped to the bottom of a 9"x13" Pyrex dish. These test tubes held the petri dishes just above the bottom of the dish, into which ~500 mL of sterile water was added. The dish was then covered with plastic wrap and tin foil and incubated at room temperature (20°C to 25°C).

Resuspension of Remaining Bacteria

At each time point, 30 ml of sterile PBS was pipetted into the petri dish completely submerging the coupon. A sterile plastic inoculating loop was used to release any viable bacteria remaining on the surface of the coating, and to mix the contents of the petri dish. Samples from this bacterial suspension were diluted in PBS, plated onto TSA plates, and incubated at 37°C for at least 24 hours. Plates containing between 30 and 300 isolated colonies were counted and used to calculate remaining bacterial concentrations.

Results and Discussion

Establishment of a Model Test System
Sterilization of the Coated Coupons

Initial studies to sterilize the coupons were undertaken to rid of any contaminant bacteria after the initial preparation of the coated coupons. Swabbing or soaking of the coupons with 70% ethanol did not completely kill bacterial surface contamination (data not shown). Further, steam sterilization (121°C, 15 psi, 15 min) disrupted the integrity of the coatings. UV irradiation of the coupons for 2 min on each side consistently killed all of the bacterial contamination without destroying the coatings.

Quantative Recovery of the Bacterial Inoculum from Coated Coupons

The bacterial strains and techniques used in this study were selected based on their use in testing disinfectants by various standard methods (36, 41, 37-40, 42). Coupons containing pigmented coatings with no antimicrobial agent (AMA) were inoculated with 0.5 ml of 10^5-10^6 CFU/ml and incubated according to the conditions outlined in the "Materials and Methods" section. Figure 1 demonstrates a nearly quantitative recovery of bacteria at all three time-points was seen. These data suggested that the bacterial cells were not tightly bound to the coating's surface and that the bacteria could be easily removed by simple washing and abrasion of the slide in PBS. Detailed experiments, however, testing the adherence of the bacteria to the coating's surface have not been performed.

Selection of Antimicrobial Agent

Initial experiments comparing the efficacy of bacterial killing by antimicrobial agents arbitrarily named "**A**" and "**B**" were performed to determine which AMA to use in further testing. The results of these experiments when using antimicrobial "**B**" demonstrated 100% killing (when compared to T_0) of both *E. coli* ATCC11229 and *S. aureus* ATCC 6538 after 12 hours. The same experiments using antimicrobial "**A**" produced a 100% reduction of *E. coli* after 12 hours but only a 68% reduction of *S. aureus* at the same time point. From these results, we selected AMA "**B**" for further use in our antimicrobial coatings.

Time Course of Antimicrobial Killing

Studies by Galeano et al and Matsumura et al (42-43) suggested that bactericidal action could be seen as early as 2 hours after bacterial contact with their antimicrobial coating. In our studies, we performed a time course looking at bacterial killing at 6 and 12 hours after the inoculation of the coated coupons. When compared to T_0, *E. coli* ATCC11229 showed 100% killing at both 6 hour and 12 hour, while *S. aureus* ATCC6538 showed 98.5% killing at 6 hour and 100% killing at 12 hour.

Dose Response with Increasing Levels of AMA

Preliminary studies were performed to determine the lowest concentration of AMA needed to kill both test organisms. In this experiment, a gradient of 0%, 0.4%, 2.0%, and 6% AMA "**B**" was used with the standard 0, 12, and 24 hour sampling times. The results in Figures 2 & 3 showed that with the coating containing 0.4% AMA "**B**", 100% killing (when compared to T_0) of *E. coli* ATCC11229 was seen at both 12 hour and 24 hour. With the same coating, 89% and >99% killing (when compared to T_0) of *S. aureus* ATCC6538 was seen at 12 hour and 24 hour, respectively. Coatings containing both 2.0% and 6.0% AMA "**B**" showed 100% killing (when compared to T_0) of the test bacteria at 12 hour and 24 hour (Figures 2 & 3).

Dose Response with Increasing Levels of Bacteria

These experiments were performed to determine the efficacy of bacterial killing with increasing levels of bacterial load. In these experiments, a standard amount of 2.0% AMA "**B**" was used with sampling at 0 hour, 12 hour, and 24 hour. Using the standard bacterial load of 10^5 CFU/ml, 100% killing was seen at both 12 hour and 24 hour (Table III). Increasing the bacterial inoculum to 10^7 CFU/ml demonstrated significant killing of *S. aureus* (>99%) at 12 hour and complete killing at 24 hour. The *E. coli* was completely killed at 12 hour and 24 hour (Table III). Using a 10^9 CFU/ml inoculum, however, required 24 hours to achieve 95% and 98% killing of *E. coli* and *S. aureus*, respectively (Table III).

Conclusions

Silver-based antimicrobial coatings have been shown to be effective in killing a variety of bacteria. In the present studies, we have demonstrated that air dry waterborne coatings containing silver-based antimicrobials effectively

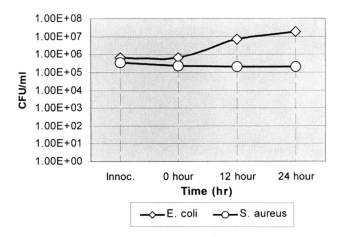

Figure 1. CFU/ml of surviving bacteria on coatings lacking any AMA (n = 7 experiments for each organism).

E. coli ATCC11229

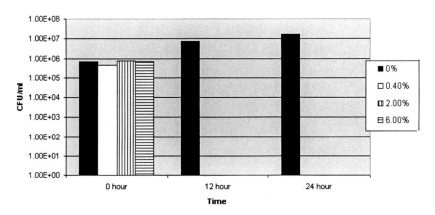

Figure 2. CFU/ml of surviving E. coli using coatings with increasing concentrations of AMA "B."

38

Table III. Dose Response using Increasing Levels of Bacteria

Time (hours)	0 hour		12 hour		24 hour	
Inoculum (CFU/ml)	E. coli[1] (%Killing)[2]	S. aureus[1] (%Killing)	E. coli (%Killing)	S. aureus (%Killing)	E. coli (%Killing)	S. aureus (%Killing)
~10^5	0	0	100	100	100	100
~10^7	0	0	100	>99	100	100
~10^9	0	0	0	0	95	98

[1] E. coli ATCC11229 and S. aureus ATCC 6538
[2] % killing was determined using the CFU/ml at T_0.

S. aureus ATCC6538

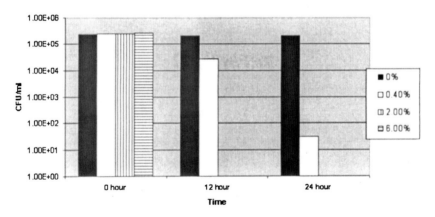

Figure 3. CFU/ml of surviving S. aureus using coatings with increasing concentrations of AMA "B."

kill the test strains of *E. coli* and *S. aureus* after 12 hour of incubation at RT. Even with increased bacterial loads (Table 3), we see $\geq$95% killing of the test organisms on the antibacterial coatings. Further experimentation using increased AMA concentrations and increased incubation times will be performed to determine if 100% killing of bacterial loads $\geq$10^9 CFU/ml can be achieved. In addition, experiments will be performed to determine the efficacy of killing with other bacteria (e.g., *Bacillus* spp., *Pseudomonas* spp., etc.).

The proposed possible applications of such coatings include areas as clinical settings, food preparation areas, and food processing facilities. The challenge with such applications is to determine the ability of such coatings to kill bacteria while having a subsequent layer of an organic nature. We are currently evaluating our antibacterial coatings containing such an organic load.

References

1. US Pat. 6,228491, Fibrous textile articles possessing enhanced antimicrobial properties.
2. Herrera, M, Carrion, p. et. al. In vitro antibacterial activity of glass-ionomer cements, Microbios, 104, 409, 2001.
3. Cho, D. L, et.al., A study on the preparation of antimicrobial biopolymer film, J.Microbiol. Biotechnol. 11(2), 193-198, 2001.
4. US. Pat Appl. 2000-57-2716, Milliken company, USA. Antimicrobial transfer substrate for textile finishing.
5. Alexander, M., K., Klibanov, J., Lewis, C., Teller, Chung-Jen Liao, Proc. Natl.Acad. Sci., 98, 5981, 2001.
6. Alexander, M. K. Klibanov, J. Lewis, C. Teller, Chung-Jen Liao, S.B. Lee, Biotechnol. Lett., 24, 801, 2002.
7. Alexander, M., K., Klibanov, J. Lewis, C. Teller, Chung-Jen Liao, S.B. Lee, Biotechol. Bioeng., 79, 466, 2002
8. Klein; M., W. F. DeGardo; Proc. Natl.Acad. Sci. 99, 5110, 2002,
9. Mackeen, C, S Person, SC Warner, W Snipes and SE Stevens, Jr. Silver-coated nylon fiber as an antibacterial agent. Antimicrob. Agents Chemother. 31, 93-99, 1987.
10. Schierholz, J.M., Z. Wachol-Drewek, L.J. Lucas and G. Pulverer. Activity of silver ions in different media. Zent.bl. Bakteriol. 287, 411-420, 1998.
11. Spadaro, A. Berger, SD Barranco, SE Chapin and RO Becker. Antibacterial effects of silver electrodes with weak direct current. Antimicrob. Agents Chemother. 6, 637-642, 1974.
12. Thibodeau , EA, SL Handelman and RE Marquis. Inhibition and killing of oral bacteria by silver ions generated with low density electric current. J. Dent. Res. 57, 922-926, 1978.

13. Klasen, H.J. Historical review of the use of silver in the treatment of burns. I. Early uses. Burns 26:117-130, 2000.
14. Klasen, H..J., A historical review of the use of silver in treatment of burns, II. Renewed interest for silver, *Burns*, 26, 131-138, 2000.
15. Maki, D.G., and P.A. Tambyah, Engineering out the risk of infection with urinary catheters. Emerg. Infect. Dis. 7:1-6, 2001.
16. Bromberg, L.E., V.M. Braman, D.M. Rothstein, P. Spacciapoli, S.M. O'Connor, E.J. Nelson, D.K. Buxton, M.S. Tonetti, and P.M. Friden. Sustained release of silver from peridontal wafers for treatment of periodontitis. J. Controlled Rel. 68:63-72, 2000.
17. Quintavalla, S., and L. Vicini. Antimicrobial food packaging in meat industry. Meat Science 62:373-380, 2002.
18. Wright, J.B., K. Lam, D. Hansen, and R.E. Burrell. Efficacy of topical silver against fungal burn wound pathogens. Am. J. Infect. Control 27:344-350, 1999.
19. von Naegelli, V. Deut. Schr. Schweiz. Naturforsch. Ges. 33, 174-182. 1893.
20. Clement, J.L. and P.S. Jarrett. Antibacterial silver. Metal-Based Drugs 1, 467-482, 1994.
21. Thurman, R.B. and C.P. Gerba. The Molecular Mechanisms of Copper and Silver Ion Disinfection of Bacteria and Viruses. CRC Critical Reviews in Environmental Control 8, 295-315, 1989.
22. Yahya, M.T., L.K. Landeen, M.C. Messina, S.M. Kute, R. Schulz, and C.P. Gerba. Disinfection of bacteria in water systems by using electrolytically generated copper:silver and reduced levels of free chlorine. Can. J. Microbiol. 36,109-116,1990.
23. Becker, R.O. Silver ions in the treatment of local infections. Metal-Based Drugs 6, 311-314,1999.
24. Carr, HS, TJ Wlodkowski and HS Rosenkranz. Silver sulfadiazine: In vitro antibacterial activity. Antimicrob. Agents Chemother. 4, 585-587, 1973.
25. Yoshida, K., M. Tanagawa, M. Atsuta. Characterization and inhibitory effect of antibacterial dental resin composites incorporating silver-supported materials. J. Biomed. Mater. Res. 47, 516-522, 1999.
26. Dibrov, P., J. Dzioba, K.K. Gosink, and C.C. Hase. 2002. Chemiosmotic mechanism of antimicrobial activity of Ag+ in Vibrio cholerae. Antimicrob. Agents Chemother. 46:2668-2670.
27. Gupta, A., L.T. Phung, D.E. Taylor, and S. Silver. 2001. Diversity of silver resistance genes in IncH incompatibility group plasmids. Microbiol. 147:3393-3402.
28. Franke, S., G. Grass, and D.H. Hies. the product of the ybdE gene of the Escherichia coli chromosome is involved in detoxification of silver ions. Microbiol. 147:965-972.
29. Nies, D.H. 1999. Microbial heavy metal resistance. Appl. Microbiol. Biotechnol. 51:730-750.

30. Gupta, A., M. Maynes, and S. Silver. 1998. Effects of halides on plasmid-mediated silver resistance in Escherichia coli. Appl. Environ. Microbiol. 64:5042-5045.
31. Silver, S., A. Gupta, K. Matsui, and J.-F. Lo. 1999. Resistance to Ag(I) cations in bacteria: environments, gene, and proteins. Metal-Based Drugs 6:315-320.
32. Slawson, R.M., E.M. Lohmeier-Vogel, H. Lee, and J.T. Trevors. 1994. Silver resistance in Pseudomonas stutzeri. BioMetals 7:30-40.
33. Deshpande, L.M., and B.A. Chopade. 1994. Plasmid mediated silver resistance in Acinetobacter baumannii. BioMetals 7:49-56.
34. Hendry, A.T., and I.O. Stewart. 1979. Silver-resistant enterobacteriaceae from hospital patients. Can. J. Microbiol. 25:915-921.
35. Barkley, W.E., and J.H. Richards. Laboratory Safety. Laboratory safety, p. 715-734 In P. Gerhardt, R.G.E. Murray, W.A. Wood, and N.R. Krieg (ed.), *Methods for General and Molecular Bacteriology*, American Society for Microbiology, Washington, D.C. 1994.
36. Luppens, S.B.I., M.W. Reij, R.W.L. van der Heijden, F.M. Rombouts, and T. Abee. Development of a standard test to assess the resistance of *Staphylococcus aureus* biofilm cells to disinfectants. Appl. Environ. Microbiol. 68, 4194-4200. 2002.
37. Rubino, J.R., J.M. Bauer, P.H. Clarke, B.B. Woodward, F.C. Porter, and H.G. Hilton. Hard surface carrier test for efficacy testing of disinfectants-collaborative study. J. AOAC Int. 75, 635-645. 1992.
38. Mariscal, A., M. Carnero-Varo, J. Gomez-Aracena, and J. fernandez-Crehuet. Development and testing of a microbiological assay to detect residual effects of disinfectant on hard surfaces. Appl. Environ. Microbiol. 65, 3717-3720. 1999.
39. Abrishami, S.H., B.D. Tall, T.J. Bruursema, P.S. Epstein, and D.B. Shah. Bacterial adherence and viability on cutting board surfaces. J. Food Safety 14, 153-172, 1994.
40. Miner, N., M. Armstrong, C.D. Carr, B. Maida, and L. Schlotfeld. Modified quantitative association of official analytical chemists sporicidal test for liquid chemical germicides. Appl. Environ. Microbiol. 63, 3304-3307, 1997.
41. Official Methods of Analysis. 15th Ed., Association of Official Analytical Chemists, Arlington, VA., Method 955.15, 961.02, 964.02, 965.19, and 966.04. 1990.
42. Galeano, B., E. Korff, and W.L. Nicholson. 2003. Inactivation of vegetative cells, but not spores, of *Bacillus anthracis. B. cereus*, and *B. subtilis* on stainless steel surfaces coated with an antimicrobial silver- and zinc-containing zeolite formulation. Appl. Environ. Microbiol. 69:4329-4331.
43. Matsumura, Y., K. Yoshikata, S. Kunisaki, and T. Tsuchido. 2003. Mode of bactericidal action of silver zeolite and its comparison with that of silver nitrate. Appl. Environ. Microbiol. 69:4278-4281.

Chapter 4

Silicone Polymers with Biocide Grafting for Antifouling/Fouling Release Coatings: Effect of Modulus on Antifouling Performance

Johnson Thomas*, Renae Fjeldheim, Seok-Bong Choi, and Philip Boudjouk

Center for Nanoscale Science and Engineering, 1805 Research Park Drive, North Dakota State University, Fargo, ND 58102
*Corresponding author: email: johnson982002@yahoo.com;
Fax: +11 701 231 5306

Modification of silicone polymers by covalently attaching organic biocides is being carried out to simultaneously incorporate antifouling/fouling release properties to silicone based marine coatings. The biocide Triclosan (5-chloro-2- (2, 4-dichlorophenoxy) phenol) was modified with alkenyl moieties and incorporated into a silicone backbone through covalent bonds. The presence of the biocide on the coating surface is expected to deter the fouling organisms from attaching to the surface of the coating. Resins were cured using vinyl terminated polydimethylsiloxane or 1, 3-cyclohexanebis methylamine depending on the crosslinking functionality on the silicone backbone. Coatings were characterized by static contact angle measurements and dynamic mechanical thermal analysis. Synthetic control over the incorporation of crosslink functionalities within the polymer resin allowed tuning of the surface of the coating and of mechanical properties. Resistance to macro fouling was tested by static immersion test in the Indian River Lagoon at the Florida Institute of Technology at different times in 2003 and 2004. Preliminary results showed that the coatings prepared from biocide incorporated silicones with the appropriate physical characteristics significantly reduced macro fouling.

43

Introduction

The release of toxic materials to the ocean water from the current metal containing marine paints is highly detrimental to marine organisms[1-4]. The issue of harbor toxicity and disposal of metal containing marine paints led to the ban on the application of environmentally harmful tributyltin (TBT)-based paint products beginning in 2003 to be completely phased out by 2008 by the International Maritime Organization (IMO) (IMO diplomatic conference, 1-5 Oct. 2001)[5]. Dissolution of copper compounds from the alternative copper ablative antifouling paints currently available is also harmful to non-targeted marine organisms[1,6]. These observations require the development of new antifouling coatings free of metal containing biocides. Most studies are now directed towards development of coatings containing non-leachable "environmentally benign" biocides fixed to a polymer in a permanent way[7-10]. This approach towards antifouling coatings is directed to the incorporation of non-metal containing organic biocides into a polymer backbone by covalent attachment to prevent the release of these biocides into the marine environment.

The aim of our research is to study siloxane polymers targeted toward the development of antifouling/fouling release coatings. Several different approaches are incorporated into this study. Cyclic methylhydrosiloxane, methylhydrosiloxane-dimethylsiloxane copolymers and polyhydromethyl siloxanes have been modified by covalently attaching organic biocides to minimize leaching. Presence of biocide on the coating surface will deter the fouling organisms from attaching to the surface or greatly reduce the attachment strength. Triclosan, which is a broad spectrum antibacterial / antimicrobial agent, was modified and covalently attached to siloxane backbones through short and long alkyl chains to explore the possibility of using this as an antifouling agent in marine coatings. In this study, triclosan is attached via covalent linkage to silicones to prevent its release into the environment. Different cross-linking moieties are used to tune the bulk properties of the coatings. The synthesis, coating preparation and fouling resistance of these coatings are reported here.

Materials and Methods

Materials

Triclosan was purchased from Lancaster Chemicals (Windham, NH). 1, 3, 5, 7- Tetramethylcyclo-tetrasiloxane (D4), 50-55% methylhydrosiloxane-dimethylsiloxane copolymer (HMS-501), polyhydromethyl siloxane (HMS 992),

Methcrylated polydimethyl siloxane (MCR-M11) and vinyldimethylsiloxy terminated polydimethyl siloxane (V 05) were purchased from Gelest (Tullytown, PA). Karstedt's catalyst (platinum (0) -1, 3-divinyl-1, 1,3,3-tetramethyl disiloxane complex), 1, 3-bis-aminomethyl-cyclohexane, allyl bromide, 1-undecenyl bromide, allyl glycidyl ether, and polybutadiene were obtained from Aldrich (Milwaukee, WI). All other reagents were obtained from Lancaster. All of the chemicals were used as received.

Experimental

Modification of Biocide by Bromo Alkene

i) In a typical reaction, allyl bromide (5g, 41mmol) was added to a solution of Triclosan (10g, 34mmol) and potassium carbonate (6g, 41mmol) in 50 ml N, N-Dimethylformamide (DMF). The mixture was stirred at room temperature overnight. After the reaction, solvent was removed by evaporation and the residue dissolved in 50ml hexane and washed with water four times (4x50ml). The organic layer was separated, dried with anh. $MgSO_4$ and evaporated to yield allyl functionalized biocide as a white solid. (10.2g, yield= 90%).
^{1}H NMR ($CDCl_3$): δ 4.58 (s, 2H), 5.2(q, 2H), 5.8(m, 1H), 6.67- 7.4 (m, 6H). ^{13}C NMR ($CDCl_3$): δ 70.09, 115.7, 117.48, 119.01, 120.42, 121.59, 122.00, 127.88, 128.43, 130.46, 132.31, 143.08, 151.34, 152.91.
ii) Triclosan was also modified by undecenyl bromide and methacryloyl chloride to obtain a long chain and ester linkage respectively.

Incorporation of Modified Biocide and allyl glycidyl ether into siloxane

i) Allyl functionalized Triclosan (4g, 12.5mmol) and allyl glycidyl ether (4.45ml, 37.5mmol) were added to a solution of D4 (3g, 12.5mmol) in 20ml of dry toluene and 2-3 drops of Karstedt's catalyst was added to the mixture and the reaction continued for 8 h at 90° C. After the reaction, the mixture was passed through neutral alumina column and solvent removed by evaporation to yield only pure product as a colorless viscous liquid (10.8g, yield = 95%).
^{1}H NMR ($CDCl_3$): δ 0.09, 0.6, 1.53, 2.57, 2.75, 3.10, 3.36, 3.41, 3.60, 3.81, 6.7-7.45. ^{13}C NMR ($CDCl_3$): δ -0.46, 13.28, 23.35, 44.54, 51.06, 71.63, 74.20, 117.48, 119.01, 120.42, 121.59, 122.00, 127.88, 128.43, 130.46, 132.31, 143.08, 151.34, 152.91.
ii) Other functionalities were also grafted on to siloxane backbone by the same procedure

Characterization of Resins

NMR spectra were recorded on a Varian V 400 MHz spectrometer.

Bulk properties of the coatings were studied using a Symyx Technologies, Inc. parallel dynamic mechanical thermal analyzer (parallel DMTA^{TM}). This is a fully parallel instrument capable of making 96 simultaneous measurements. Modulus measurements were performed by measuring the force needed to deform a thin polyimide substrate by a given amount with and without a sample present. The method and theory of this instrument is documented by Kossuth *et al.*[11]. Samples were deposited on standard DMTA plates supplied by Symyx Technologies (Santa Clara, CA). Coating samples were prepared by depositing 20-25µl of the resin mixed with the appropriate crosslinker on to regions 5mm diameter on the DMTA plate. Four replicates were deposited on the plate for each sample. After the samples were deposited, the DMTA plate was kept in an oven and heated at 60°C for 48 h to completely cure all the coating samples. After curing, the thickness of each sample was determined using a laser profilometer. The height profile of each sample was recorded and fit to a square cross section and the height of the fitted profile was taken as the thickness of the film. The measured thicknesses range from 250 µm -300 µm. After the thickness measurements, the sample plate was introduced into the parallel DMTA and measurements were taken. These experiments were carried out in the temperature range -125°C to 150° C with a ramp rate 5°C/min. at 10 Hz.

Preparation of Coatings

Table I show a list of coatings used for this study. They were prepared mainly by two methods. Glycidyl ether functionalized resins were mixed with 1,3-cyclohexane-bis (methylamine) (1epoxy equivalent /1amine equivalent) and applied on aluminum panels. The coatings were generally touch dry in 3 h and were further cured at 60°C for 24 h. Resins having residual Si-H groups were cured by hydrosilation using divinyl terminated polydimethyl siloxane and polybutadiene (1.2 equivalent SiH/ 1 equivalent double bond). These coatings were generally touch dry in 6 h and were further cured at 60° C for 48 h. The dry film thicknesses of the coatings range from 250µm to 300µm.

Panel Preparation and Deployment Site

Coatings were applied on 4" x 8" marine grade aluminum panels. The panels were cleaned and roughed with sand paper (400 grits) followed by application of the anti corrosive epoxy primer coating, Macropoxy 646, from

Table I. List of coatings and their chemical characteristics

Coating No.	Matrix	Biocide	Crosslinking group	MCR-M11
TJ01-A	HMS-501	1 Allyl	7 Epoxy	-
TJ01-B	HMS-501	2 Allyl	6 Epoxy	-
TJ01-C	HMS-501	3 Allyl	5 Epoxy	-
TJ01-D	HMS-501	4 Allyl	4 Epoxy	-
TJ02	HMS-992	8 Undecyl	6 Epoxy	-
TJ03	HMS-992	8 Allyl	Epoxy	2
TJ04	HMS-992	8 Undecyl	6 -Si-H	-
TJ05	HMS-992	8 Undecenyl	6 -Si-H	2
TJ07	HMS-992	8 Undecenyl	6 -Si-H	2
TJ08	HMS-501	4 Methacrylate	4 Epoxy	-
TJ09	HMS-992	8 Allyl	6 Epoxy	-
TJ10	HMS-992	8 Undecenyl	10 -Si-H	-
TJ11	HMS-992	8 Undecenyl	10 -Si-H	2

HMS-501- 50-55% Methyl hydrosiloxane- Dimethyl siloxane copolymer (Mn= 900-1200), HMS-992- polyhydromethyl siloxane (Mn= 2000), MCR-M11-MonoMethacryloxypropyl Terminated polydimethyl siloxane (Mn=800-1000).

Sherwin Williams using airless spray equipment. All the experimental coatings were applied on top of the epoxy primer layer without any tie-coat, by a draw down bar.

HMS 501 (Reference) with out any modification was cured with divinyl siloxane and was used as a control coating so that the antifouling property of biocide incorporated HMS 501 could be compared. Intersleek 425 (Fouling release coating from International Paints, U.K.) and Copper ablative coating BRA 642 were also used as a reference to compare the antifouling/fouling release properties of the experimental coatings.

Static immersion tests were carried out at Florida Institute of Technology in the Indian River Lagoon (Melbourne, FL). Before immersion were subjected to

15 day pre-leaching in artificial seawater (Aquarium Systems Inc., Sarrebourg, France) under static conditions, to remove any unreacted materials including unreacted biocide from the coatings. The absence of unreacted biocide was confirmed by HPLC analysis. Water was changed after each five-day period during pre-leaching.

All panels were held one meter below the water surface inside ½" galvanized mesh cages. This was done to protect the panels from fish and other aquatic organisms other than the foulants. Four replicates of each coating were used for the immersion studies and the fouling rating of the coatings were calculated based on ASTM fouling rating. Fouling rating (FR) is defined by FR= 100- the sum of fouling % cover (not including slime), i.e., FR 100 is a surface free of macro fouling. The static immersion studies were carried out at different times in 2003 and 2004.

Results & Discussion

Synthesis and Chemical Characterization

The aim of this study was to survey different approaches within the synthetic scheme for obtaining antifouling / fouling release coatings. An organic biocide, Triclosan, was incorporated into the siloxane backbone via covalent linkage for imparting antifouling properties. Siloxanes are known to be low modulus materials, which promote foul release. It was expected that attaching an organic biocide to a polysiloxane would result in an antifouling material with possible fouling release properties. Allyl bromide (short chain) and 1-undecenyl bromide (long chain) were used to incorporate the biocide into siloxane. The long alkyl chain was selected to provide more flexibility so that the biocide will be more available on the top layer of the coating providing an antifouling surface.

Triclosan-alkenyl derivatives were incorporated into the siloxane backbone by hydrosilylation using Karstedt's catalyst[12]. The modification of the biocide and its incorporation into linear polydimethylsiloxane-co-methylhydrosiloxane and its further modification by allyl glycidyl ether are given in Fig. 1.

First, Triclosan was modified by alkenyl bromide to facilitate hydrosilylation. In one set of coatings, this biocide incorporated resin (I) with residual Si-H groups was used to prepare coatings using vinyl terminated PDMS or polybutadiene as the crosslinker. In another set of experiments, biocide incorporated siloxanes were further modified by the incorporation of allyl glycidyl ether (II) to add another crosslinking type. These glycidyl ether modified resins were crosslinked with bisamine.

Effect of Modulus and Biocide loading on Fouling Rating

From our initial studies[13] we found that the fouling rating was strongly dependent on the modulus of the coatings. To study this systematically, we varied the modulus as well as biocide loading in the coatings. As shown in Table I, the number of cross linking groups in the coatings from TJ01-A to TJ01-D decreases and this leads to a corresponding decrease in the modulus of these coatings (Figure 2). A few coatings were also prepared by grafting silicone fluid on to the siloxane backbone to further soften the coating. The static immersion studies of these coatings were carried out between April 26, 2004 and July 16, 2004. Two inspections were done on these coatings, one after 44 days and another one after 80 days of immersion. The fouling rating of these coatings after 44 days and 80 days static immersion studies at FIT are given in figure 3 and figure 4 respectively. It is interesting to note that at the first inspection, the fouling rating was strongly dependent on the modulus of the coatings. When the modulus decreased within the same set of coatings (TJ01-A to TJ01-D), the fouling rating showed a marked increase. The panel pictures in figure 5 show that there is significant reduction in fouling on the coatings as the modulus decreases. However, increase in biocide loading or grafting of silicone fluid did not change the FR significantly. Statistically the coatings have similar FR.

Second inspection of the coatings after 80 days showed that all the coatings have lower FR except for TJ05, which contained silicone fluid grafted on the silicone backbone. The reason for the low fouling rating lies in the fact that all the coatings are soft silicone coatings and barnacles cut through the coatings and grow on the epoxy primer applied to the aluminum panels. Added to this is the poor adhesion between the top coating and the primer, which leads to blistering and delamination from the primer (Figure 6). The combined biological corrosion and the poor adhesion between the primer and the top coating affect the durability of the coatings and they tend to foul after a period of eight weeks.

Conclusion

The biocide Triclosan has been chemically tethered to silicone coatings for antifouling / fouling release applications. This study shows that the bulk properties of the coatings can be significantly varied by the nature of the crosslink and the nature of side groups. Further, the modulus of the coatings can be tuned by varying the number of cross-linking functionality on the backbone. Their resistance to macro fouling was evaluated by static immersion tests in the Indian River Lagoon at Florida Institute of Technology. The static immersion study indicates that the coatings with covalently linked biocide and appropriate physical characteristics are effective in preventing macro fouling. The bulk properties such as Tg and modulus of the coating tend to strongly influence the

50

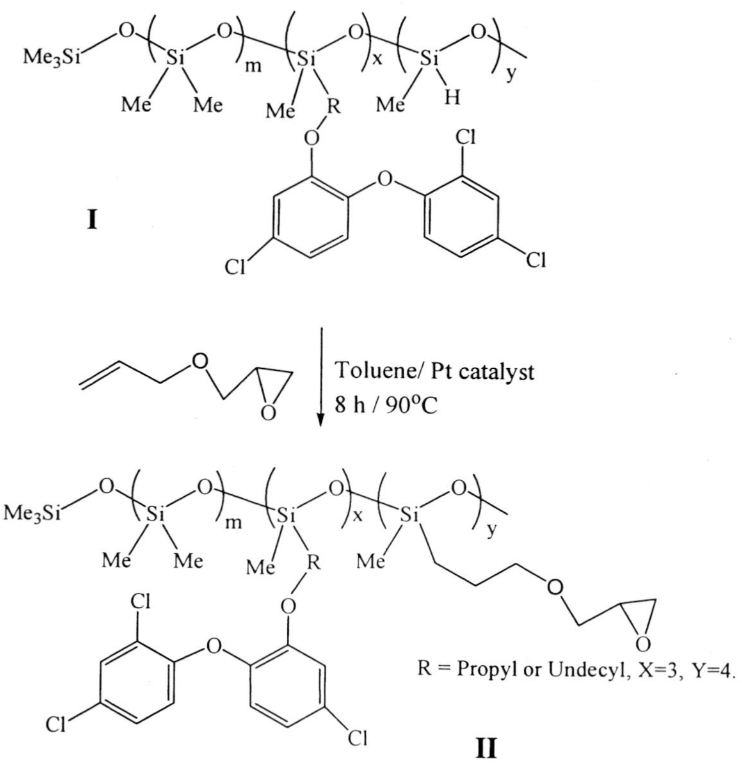

Figure 1. Overall synthetic scheme showing the modification of biocide and incorporation of modified biocide and glycidyl ether into siloxane.

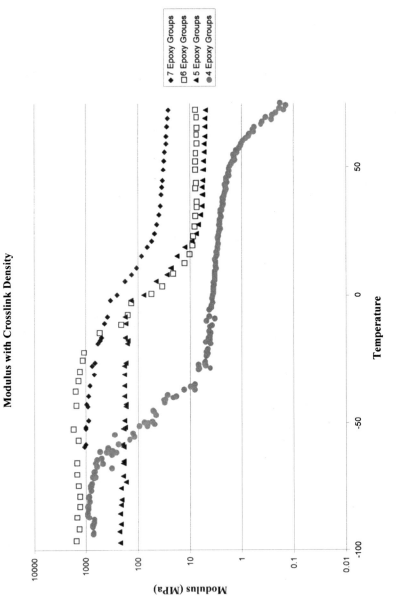

Figure 2. DMTA plots of the coatings showing the variation of storage modulus with temperature

53

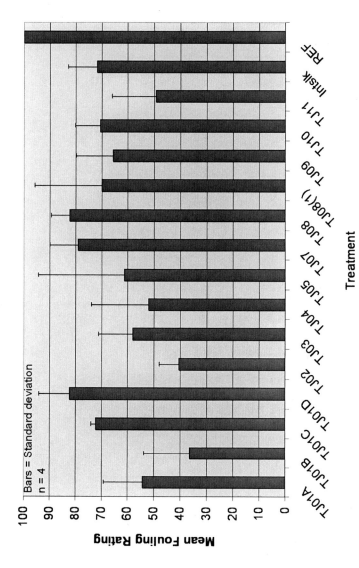

Figure 3. Average fouling rating of the coatings after 44 days static immersion experiments in the Indian river lagoon at Florida Institute of Technology.

54

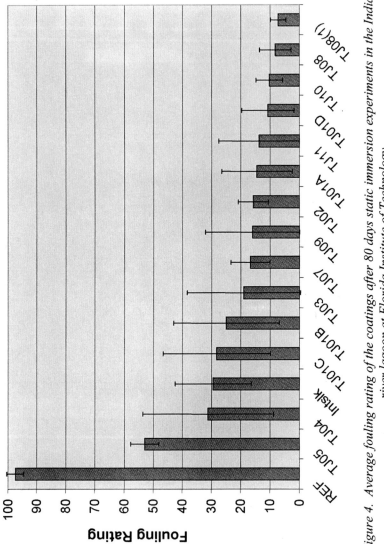

Figure 4. Average fouling rating of the coatings after 80 days static immersion experiments in the Indian river lagoon at Florida Institute of Technology.

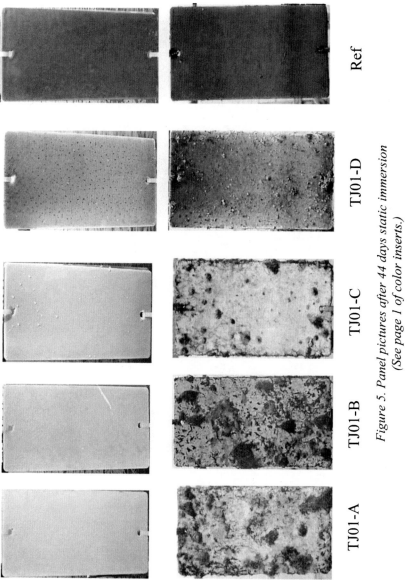

Figure 5. Panel pictures after 44 days static immersion (See page 1 of color inserts.)

Ref

TJ01-D

TJ01-C

TJ01-B

TJ01-A

56

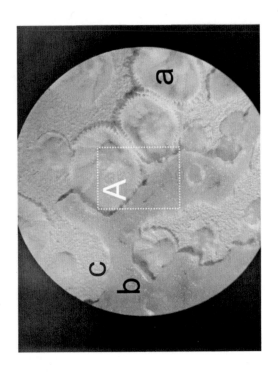

a) Barnacle basis
b) Experimental coating
c) Undercoating

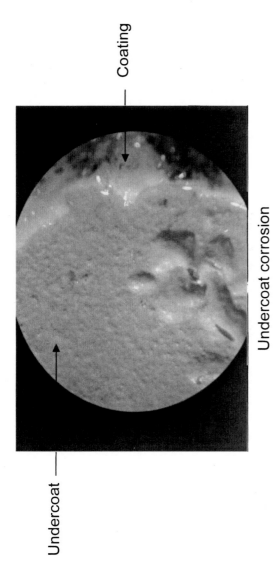

Coating

Undercoat

Undercoat corrosion

Figure 6. Barnacle cutting and corrosion in the coatings.
(See page 2 of color inserts.)

fouling resistance. High modulus and very low modulus coatings show high macro fouling. In the case of high modulus coatings, biocides may be prevented from migrating to the surface of the coatings by the highly cross-linked network and therefore they are not effective in providing a biocidal surface. In the case of low modulus coatings, they fail very early and delaminate from the tie coat and the fouling organisms grow on the tie coat. These coatings fail in the long term due to the softness as well as poor adhesion to the primer.

Acknowledgment

Financial support from the Office of Naval Research through ONR grants # N00014-02-1-0794 and N00014-03-1-0702 are gratefully acknowledged. Authors are thankful to Prof. Geoffrey Swain and Mr. Kris Kavanagh at the Florida Institute of Technology (FIT), Melbourne, FL, for conducting the static immersion studies. The assistance of Jim Bahr and Christine Graewalk at Center for Nanoscale Science and Engineering (CNSE) for conducting the DMTA experiments is highly acknowledged.

References

1. Katranitsas, A.; Castritsi-Catharios, J.; Persoone, G. *Marine Pollution Bulletin* **2003**, 46(11), 1491-1494.
2. Alzieu, C. *Ecotoxicolog,y* **2000**, 9(1/2), 71-76.
3. Claisse, D; Alzieu, C. *Marine Pollution Bulletin* **1993**, 26(7), 395-7.
4. Evans, S. M. *Biofouling* **1999**, 14(2), 117-129.
5. Champ, M. A. *Marine Pollution Bulletin* **2003**, 46(8), 935-940.
6. Terlizzi, A.; Fraschetti, S.; Gianguzza, P.; Faimali, M.; Boero, F. *Aquat Conserv Mar Freshw Ecosyst* **2001**, 11, 311-317.
7. Ikeda, T.; Yamaguchi, H.; Tazuke, S. *Antimicrobial Agents and Chemotherapy 1984*, 26(2), 139-44.
8. Hazziza-Laskar, J.; Helary, G.; Sauvet, G. *Journal of Applied Polymer Science 1995*, 58(1), 77-84.
9. Nurdin, N.; Helary, G.; Sauvet, G. *Journal of Applied Polymer Science* 1993, 50(4), 663-70.
10. Sauvet, G.; Dupond, S.; Kazmierski, K.; Chojnowski, J. *Journal of Applied Polymer Science 2000*, 75(8), 1005-1012.

11. Kossuth, M. B.; Hajduk, D. A.; Freitag, C.; Varni, J. *Macromolecular Rapid Communications* 2004, 25, 243-248.
12. Chauhan, M.; Hauck, B. J.; Keller, L. P.; Boudjouk, P. *Journal of Organometallic Chemistry* **2002**, 645(1-2), 1-13.
13. Thomas, J.; Choi, S.B.; Fjeldheim, R.; and Philip Boudjouk, P. *Biofouling,* **2004**, 20(4/5), 227-236.

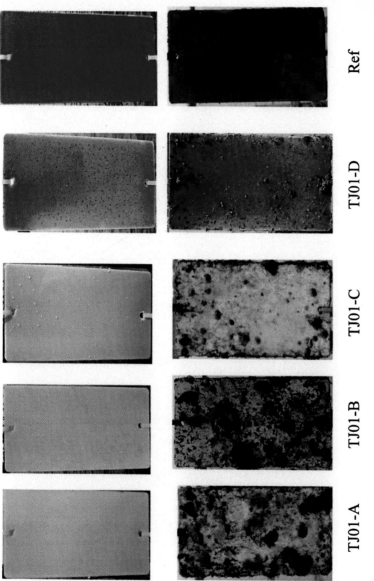

Figure 4.5. Panel pictures after 44 days static immersion

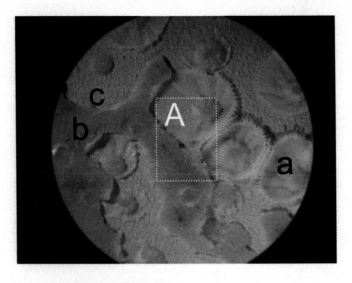

a) Barnacle basis
a) Experimental coating
b) Undercoating

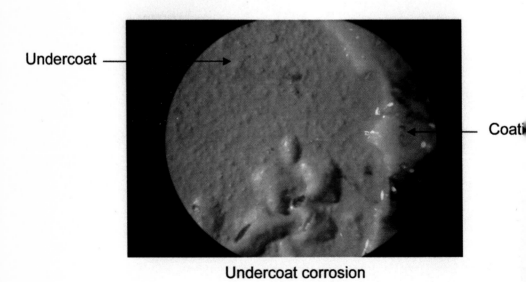

Undercoat

Coati

Undercoat corrosion

Figure 4.6. Barnacle cutting and corrosion in the coatings.

Figure 10.8. Time = 0 hours   *Figure 10.9. Time = 336 hours*

(Reproduced with permission from Reference 24. Copyright 2004
Rapra Technology Ltd.)

Figure 10.10. BAM-PPV + MIL-P-53022 + MIL-PRF-85285

Figure 10.11. BA M-PPV ± MJL-PRF-23377C + MJL-PRF-85285

Figure 10.12. CCC + MIL-PRF-23377C + MIL-PRF-85285

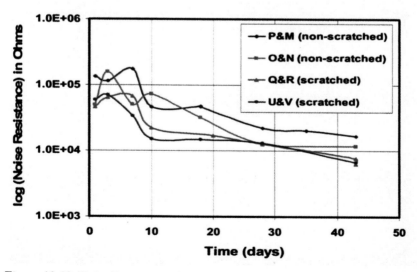

*Figure 10.13. Noise Resistance (Ω, log scale,) of BAM-PPV-coated Al 2024-T3
as a function of time of immersion*

Chapter 5

Thermoset Siloxane–Urethane Fouling Release Coatings

Partha Majumdar, Abdullah Ekin, and Dean C. Webster*

Department of Coatings and Polymeric Materials, Center for Nanoscale Science and Engineering, North Dakota State University, Fargo, ND 58105

The exploration of numerous variables involved in the design of novel crosslinked siloxane-urethane coatings was carried out using combinatorial methods. Libraries of coatings that explored the effect of siloxane level, solvent, and other variables were prepared and characterized for their surface energy and pseudobarnacle pull-off adhesion. Stability of the coatings was determined by measuring the surface energy after aging the coatings in water and re-meauring the surface energy. Coating compositions and formulations were identified that had stable hydrophobic surfaces and also had low pseudobarnacle adhesion values.

Fouling of ship hulls by marine organisms has plagued shipping activities for millennia. Fouling presents a significant drag penalty, reducing the efficiency of propulsion and resulting in the expenditure of excess fuel to overcome the drag in order to meet target cruising speed. Fouling also results in the transport of organisms from foreign ecosystems and can result in the introduction of invasive species. Thus, from both an operational and ecological point of view, methods to reduce or eliminate fouling are necessary.

While a number of approaches have been taken to prevent fouling, the most successful approach has been the use of coatings containing biocidal chemicals. Organisms are simply killed and either do not settle or are easily removed. Organo-tin and copper compounds have been in use since the 1970s. Although highly effective at reducing fouling, these biocidal agents have been linked to environmental problems. Release of the compounds from the coatings has led to sediment accumulations of the toxins resulting in harm to non-targeted sea life (i.e. oysters) (1-6). The International Maritime Organization has proposed a ban on new applications of organo-tin coatings starting in 2003 and complete removal of these coatings from all ships by 2008. While organo-tin compounds will initially be replaced by other less toxic biocides, such as copper or organic biocides, coating systems that do not leach any kind of biocidal compounds are desired.

Fouling release (FR) coatings appear to be a leading non-toxic alternative to biocide containing coatings. These are coating systems that do not necessarily prevent the settlement of marine organisms, but permit their easy removal with the application of shear to the surface of the coating. Ideally, the hydrodynamic shear on the hull as a ship reaches cruising velocity be sufficient to remove fouling organisms; however, coatings that are also easily cleaned is also a desired property of the coating system. The most successful of these coatings are based on silicone elastomers (7-9). It has been shown that coatings with low modulus and low surface energy provide easy release of fouling organisms (10). However, due to their low modulus, these coatings are easily damaged. In addition, they may also suffer from poor adhesion, poor durability, and high cost.

To provide coatings that exhibit fouling release behavior while also yielding improved durability, self-stratifying coatings that phase separate into a low surface energy, low modulus top layer, with a tougher lower layer can be designed. Self-stratifying coatings are coatings that are applied in a single step, but then spontaneously phase separate into two or more distinct layers upon application and film formation. Surface energy and viscosity are the main driving and/or controlling forces for self-stratification (11-15). A coating composed of poly(dimethyl siloxane) (PDMS) and polyurethane components may meet these requirements. Since surface energy is a primary driving force, the PDMS component will form the top, low energy, rubbery layer. The polyurethane component will form the tough durable underlayer. An additional advantage of this system is that the isocyanate resins used to form the polyurethane may react with residual hydroxyl groups on the epoxy anti-corrosion primer, providing good adhesion, thus eliminating the need for a tie layer between the corrosion coating and the fouling-release coating, resulting in good adhesion.

Siloxane-polyurethane block and graft copolymers have been evaluated for use as anti-fouling coating systems (*16-21*). While these systems had good initial fouling release behavior, performance degraded significantly after prolonged exposure to water. This was found to be due to migration of the hydrophilic hard block segments to the surface, resulting in a change in surface energy (*22-25*). These systems, however, were thermoplastic coatings, while our approach results in *crosslinked* coatings. The presence of crosslinking is expected to lock the structure in place and result in coatings that do not rearrange on exposure to water.

Figure 1 illustrates two conceptual possibilities for self-stratification: the system may phase separate into two discernable layers, or a gradient in composition may exist between the polyurethane and polydimethyl siloxane layers. In either case the PDMS is covalently attached to the overall polyurethane coating. The actual situation in these systems is probably more complex than either picture.

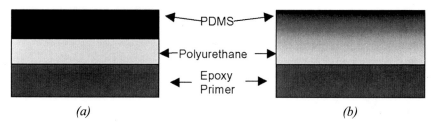

PDMS

Polyurethane

Epoxy
Primer

(a)   *(b)*

Figure 1. Possible cross-section morphology of self-stratifying coating systems illustrating (a) distinct phase separation and (b) a gradient in composition.

In these coatings the polymer system will consist of an organofunctional PDMS, an organic polyol, and a polyisocyanate crosslinker. A large number of variables are expected to influence the development of the stratified structure, the thickness of the PDMS layer, and the toughness of the coating system. For example, the molecular weight of the PDMS segment may determine the thickness of the top layer of the coating. The amount of PDMS relative to the other components is also expected to play a role. The reactive end group (hydroxyl alkyl, amino alkyl) of the PDMS will influence its reactivity with the isocyanate crosslinker. The composition and functionality of the polyol and the isocyanate used in the polyurethane component will affect the modulus and crosslink density of the coating system. A number of additional formulation variables are also expected to play a significant role. The solvent composition can affect the compatibility of the various components initially and influence the compatibility as the coating cures. Finally, the cure speed, compatibility of the oligomers, and the solvent evaporation rates will all operate simultaneously to influence the development of the network and the two-phase structure of the coating.

High throughput methods can provide a means to explore this large variable space to ensure that compositions that have a unique combination of performance properties are not missed (*26-28*). In this work, we describe the preparation of several libraries of siloxane-urethane coatings using automated systems and screening of these coatings for their initial surface properties and change in properties after water exposure in order to determine the stability of the coatings. In addition, the pseudobarnacle adhesion properties of several coating libraries are also explored.

Experimental

Materials

Solvents methyl n-amyl ketone (MAK), butyl acetate (BA), and ethyl 3-ethoxy propionate (EEP) were obtained from Eastman Chemical Company. Aromatic 100, toluene, and mixed xylenes, dibutyl tin diacetate (DBTDA), and 2-4-pentanedione (PDO), were obtained from Aldrich. Polycaprolactone triol (Tone 0305, PCL) was obtained from Dow Chemical. Polyisocyanates Tolonate XIDT 70B and HDT 90 were obtained from Rhodia. These are isocyanaurate trimers of isophorone diisocyanate and hexamethylene diisocyanate, respectively. Desmodur N3600 was obtained from Bayer; this is a trimer of hexamethylene diisocyanate with a narrow molecular weight distribution. Octamethylcyclotetrasiloxane (D4) and 1,3-bis(3-aminopropyl)-1,1,3,3-tetramethyldisiloxane were obtained from Gelest. The hydroxyethyloxypropyl PDMS of MW=10,000 g/mole was obtained from Chisso. The structures of the materials used are shown in Figure 2.

Synthesis of aminopropyl terminated polydimethyl siloxane

The synthesis of an aminopropyl terminated polydimethyl siloxane (PSX-NH2) of amine equivalent weight 6,000 (~12,000 MW) was carried out as follows: In a 250-ml three necked round bottom flask 2.04 g 1,3-bis(3-aminopropyl)-1,1,3,3-tetramethyldisiloxane and 10 g D4 were mixed. The solution was heated with stirring under nitrogen. When the temperature reached 80°C, 0.1% catalyst (tetramethylammonium 3-aminopropyl dimethyl silanolate) (*29*) was added. After one hour of heating the viscosity increased slightly, and the remaining D4 (90 g) was placed into an addition funnel and added dropwise to the solution. After the completion of the addition of all of the D4 (7-10 hours), the heating was continued for an additional 2-3 hours. Then, the temperature was increased from 80°C to 150°C and kept at that temperature to decompose the catalyst. After decomposition, the reaction mixture was cooled to room temperature. The PDMS polymer having an amine equivalent weight of 12,000 g/eq ~24,000 MW) was prepared using a similar procedure, but half of the amount of disiloxane was used.

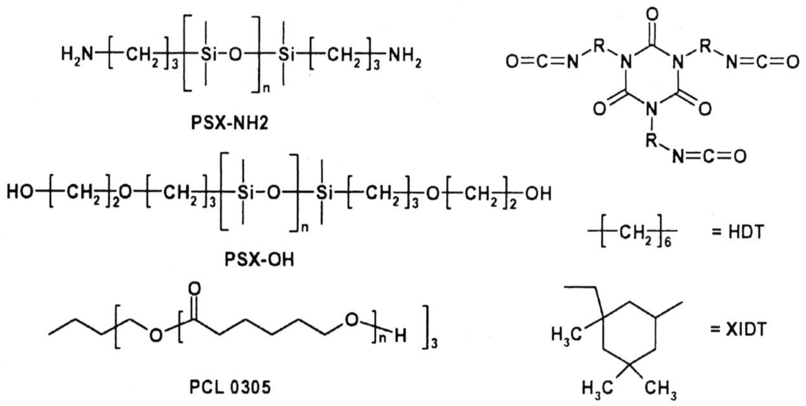

Figure 2. Structures of the reactants used in the coatings.

Formulation and Application

Formulations were prepared using a Symyx automated formulation system (*28*). Libraries were designed using Symyx Library Studio software. All libraries consisted of 24 compositions in a 4 x 6 array format. Isocyanate to hydroxyl (or hydroxyl plus amine) ratio was 1.1:1 in all cases. All libraries used PCL 0305 as the organic polyol. The PDMS, isocyanate used, and key variables for each library evaluated are summarized in Table I. Catalyst, DBTDA, was 0.01 weight percent based on resin solids. Ten percent 2,4-pentanedione was used in all formulations as a pot life extender. All reagents except for the crosslinker were dispensed to the entire array according to the design and mixed with magnetic stirring. The crosslinker was then dispensed to each vial and mixed. The library was then transferred to the Symyx coating application system and 24 coatings drawn down in array format over aluminum test panels. Coatings were cured in an oven at 80°C for 1 hour. Film thickness is approximately 25-50 μm.

Coating Screening

Coatings were screened for initial surface energy using an automated contact angle system (*28*). The water contact angle (CA) data reported is the

Table I. Summary of the libraries prepared

Library	PDMS (MW)	Isocyanate	Row	Column
A	PSX-NH2 (24K)	XIDT	Solvent	% PDMS
B	PSX-NH2 (24K)	HDT	Solvent	% PDMS
C	PSX-NH2 (12K)	XIDT; HDT	Time	Mixed: Solvent, Isocyanate
D	PSX-OH (10K)	N3600	% PDMS	Solvent

average of three measurements. To check for surface stability, coating panels were aged in deionized water, the coatings were dried and surface energy measurements were made. Coatings were analyzed for pseudo-barnacle adhesion (*10*) using a Symyx automated adhesion tester. In this test, aluminum studs are attached to the coating samples using an epoxy adhesive (Hysol Epoxy Patch 1C). Following curing, an automated pull-off device pulls the stud and measures the maximum force at release. Three adhesion tests are conducted per coating sample.

Results and Discussion

In order to identify a siloxane-urethane composition that has a suitable combination of properties, a number of variables must be explored over a wide range. The high throughput approach is a methodology that can be used to accelerate this process, allowing us to screen a large number of variables in a single experiment.

In these experiments, we are interested in exploring several variables that will lead to a stable hydrophobic surface, determining the effect of the amount of PDMS on the surface properties, and also identifying key variables that affect the release properties of the coatings. To this end, we prepared a series of combinatorial libraries to explore these variables.

In preliminary experiments, to establish starting points for more detailed experimentation, we explored the use of several different PDMS polymers having reactive endgroups (aminopropyl, hydroxyl propyl, etc.), several commercially available polyols, catalysts, solvent mixtures, and the use of a pot life extender. In a combinatorial experiment, having a suitable pot life for the coating formulations is important since the first coating mixed must not gel before the last coating in a library is mixed (*30*). 2,4-pentanedione was found to function as an effective pot life extender for these coatings. Coatings prepared both with and without the pot life extender in the laboratory had the same physical and mechanical properties.

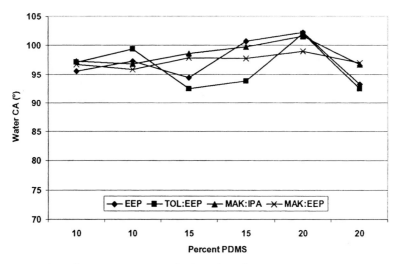

Figure 3. Initial water contact angle data for Library A. Data series represent solvent composition of the coating formulations.

Poly(dimethyl siloxane) has a much lower solubility parameter than the other oligomers and components used in the formulation of these coatings. Thus, we found that finding a solvent blend that would help compatibilize all of the coating components in solution was challenging, but necessary. With some solvent mixtures, as soon as the agitation was stopped, the coating formulation would phase separate into two distinct layers. Attempts at making coatings from this kind of unstable mixture usually resulted in gross phase separation of the PDMS to the surface of the coating such that the PDMS had not reacted with the isocyanate and so was not chemically incorporated.

Library A was designed using an aminopropyl terminated PDMS with the polycaprolactone triol as the organic polyol and IPDI trimer (XIDT) as the crosslinker. Since amines react more readily with isocyanate than hydroxyls, it was thought that using an aminopropyl terminated PDMS would help ensure that the PDMS was incorporated into the polyurethane network. The amine reactive groups could begin reacting with the isocyanate after the coating was mixed. The PDMS range was 10-20 percent. In addition, we wanted to explore the effect of solvent on the properties of the coatings and formulated the system using several different solvent compositions. The initial water contact angle data for the library is given in Figure 3. All of the coatings prepared in this experiment are initially hydrophobic (CA>90).

The contact angle data and change in contact angle after 30 days of water immersion for this library is given in Figure 4. While most of the coatings had a decrease in contact angle, the decrease in contact angle is not excessive and many of the coatings remained hydrophobic. In contrast, several of the coatings that had used a mixture of toluene and EEP as the solvent blend became slightly more hydrophobic.

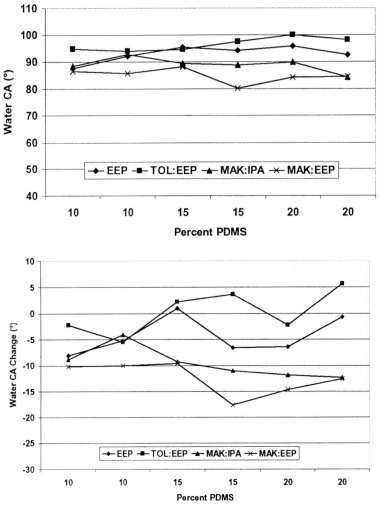

Figure 4. Water contact angle and contact angle change of Library A after immersion in deionized water for 30 days.

The automated pull-off adhesion system was used to determine the pseudobarnacle adhesion of this series of coatings. This test measures the force required to remove an epoxy from the surface of the coating, and may be indicative of how strongly a barnacle would adhere to the coating. In Figure 5 a strong trend is observed depending on the solvent used with the coatings made using EEP as the solvent having the lowest release force. We believe this to be a manifestation of the morphology of the coating system that is developed during the curing process. Consider the fact that this is a highly dynamic system once the crosslinker is mixed into the coating formulation. At this stage the isocyanate can begin to react with the polyol and the PDMS. Following application to the substrate, the low surface energy PDMS will tend to migrate to the air interface (the coating surface) and also phase separate from the polyurethane component. While the initial solvent blend helps compatibilize the components, as the solvents evaporate at different rates from the coating, the compatibility can also be changing. All of this can contribute to the extent of phase separation and result in different self-stratified morphologies. This variation in morphology is reflected in the pull-off adhesion data. The lowest pull-off adhesion value is achieved when the slowest evaporating solvent is used in the system, EEP, followed by coatings prepared using a mixture of EEP and toluene.

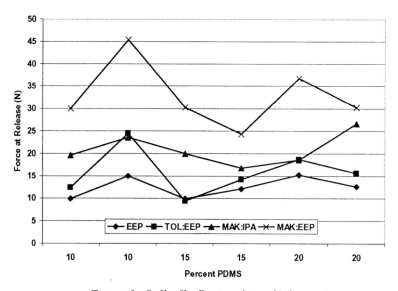

Figure 5. Pull-off adhesion data of Library A.

A similar library to the previous one was prepared with the exception that the crosslinker was a triisocyanurate of hexamethylene diisocyanate (HDT) (Library B). The initial contact angle data is shown in Figure 6 and indicates that, with a few exceptions, most of these coatings are initially hydrophobic. The pseudobarnacle pull-off adhesion again shows a dramatic effect of the solvent used (Figure 7). The coatings made using either EEP or a combination of toluene and EEP had the lowest removal force.

A lower molecular weight aminopropyl PDMS was used to prepare Library C and this library was designed to survey a range of PDMS levels (10, 15, 20 percent) using the two isocyanate crosslinkers and the two best solvent systems.

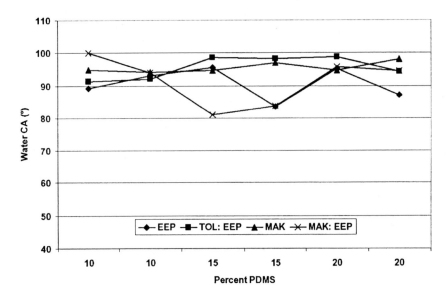

Figure 6. Water contact angle data for Library B.

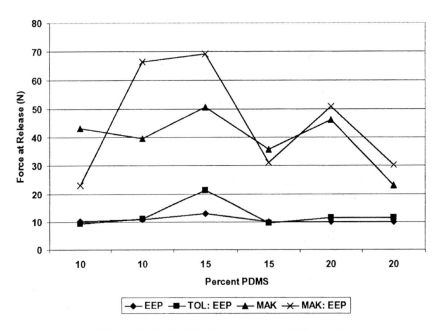

Figure 7. Pull-off adhesion data for Library B.

The water contact angle was determined weekly for four weeks. As can be seen in Figure 8, the coatings were all hydrophobic and remained hydrophobic during the testing period. This indicates that these coatings do not undergo significant reorganization in an aqueous environment and maintain their hydrophobic surfaces.

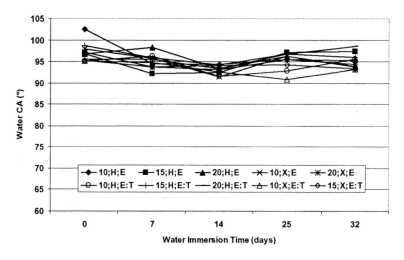

Figure 8. Water contact angle data for Library C. Data series are given as Percent PDMS:Isocyanate (H=HDT; X=XIDT):Solvent(E=EEP,E:T=EEP and Toluene blend)

Finally, a library was prepared using two levels of a hydroxyl alkyl terminated PDMS with Desmodur N3600 as the crosslinker and using several different solvent blends. The solvent blends used are described in Table II. The initial water contact angle data is shown in Figure 9. While most of the coatings were hydrophobic, several coatings were hydrophilic. In particular, most of the coatings based on a blend of IPA, EEP, and MAK were hydrophilic initially.

This set of coatings was aged in water for 30 and 60 days and the 60 day water contact angle and change in contact angle data are presented in Figure 10. The coatings were generally stable with very little change in contact angle following exposure. Several of the coatings had an increase in contact angle, including three coatings that were slightly hydrophilic initially. The coatings based on 20% PDMS and the IPA:EEP:MAK solvent blend had contact angles below 90°. Thus, in this system, a variety of solvents can be used to prepare coatings that are hydrophobic and remain hydrophobic after prolonged exposure to water.

72

Table II: Solvent combinations used in Library D.

Designation	BA	EEP	Toluene	IPA	BA	MAK
B:E:T	33.3	33.3	33.3			
I:T			50	50		
I:B	50		50			
I:B:T	27		4	69		
I:M				66		34
I:E:M		5		68		27

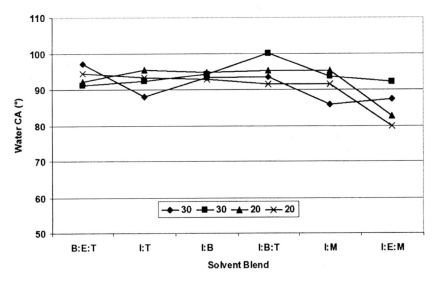

Figure 9. Water contact angle data for Library D. Data series is percent PDMS.

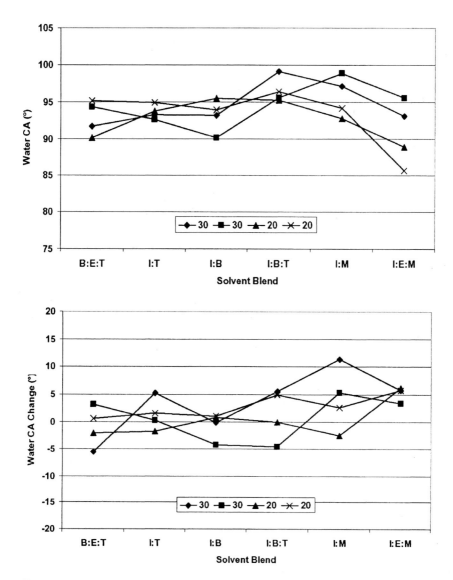

Figure 10. Water contact angle and change in water contact angle for Library D after immersion in deionized water for 60 days.

Conclusions

In a series of combinatorial experiments, the effect of key composition and formulation variables on the formation of siloxane-urethane coatings was explored. Combinatorial methods are an efficient means for exploring this large variable space. In this study, 82 compositions/formulations were studied in only four experiments. Coatings that used an aminopropyl terminated PDMS were generally stable following water immersion. The solvent composition was shown to plays an important role in the formation of stable coatings and also appears to contribute significantly to the morphology of the coatings according to the variations observed in the pseudobarnacle adhesion tests.

Acknowledgements

We would like to acknowledge support from the Office of Naval Research under grants N00014-03-1-0702, and N00014-04-1-0597.

References

1. C. Alzieu. *Ocean & Coastal Management* **1992**, *40*, 23.
2. M. A. Champ. *The Science of the Total Environment* **2000**, *258*, 21.
3. P. McClellan-Green; J. Robbins. *Marine Environmental Research* **2000**, *50*, 243.
4. S. Nehring. *Journal of Sea Research* **2000**, *43*, 151.
5. M. G. Marin; V. Moschino; F. Cina; C. Celli. *Marine Environmental Research* **2000**, *50*, 231.
6. A. P. Negri; L. D. Smith; N. S. Webster; A. J. Heyward. *Marine Pollution Report* **2002**, *44*, 111.
7. R. F. Brady, Jr.; I. L. Singer. *Biofouling* **2000**, *15*, 73-81.
8. R. F. Brady. *Progress in Organic Coatings* **1999**, *35*, 31-35.
9. R. F. Brady. *Progress in Organic Coatings* **2001**, *43*, 188-192.
10. J. G. Kohl; I. L. Singer. *Progress in Organic Coatings* **1999**, *36*, 15-20.
11. V. Verkholantsev; M. Flavian. *Progress in Organic Coatings* **1996**, *29*, 239-246.
12. P. Vink; T. L. Bots. *Progress in Organic Coatings* **1996**, *28*, 173-181.
13. S. Benjamin; C. Carr; D. J. Walbridge. *Progress in Organic Coatings* **1996**, *28*, 197-207.
14. D. J. Walbridge. *Progress in Organic Coatings* **1996**, *28*, 155-159.
15. C. Carr; E. Wallstoem. *Progress in Organic Coatings* **1996**, *28*, 161-171.
16. T. Ho; K. J. Wynne; R. A. Nissan. *Macromolecules* **1993**, *26*, 7029-7036.

17. J. Chen; J. A. Gardella, Jr. *Macromolecules* **1998**, *31*, 9328-9336.
18. H. Zhuang; J. A. Gardella, Jr.; D. M. Hercules. *Macromolecules* **1997**, *30*, 1153-1157.
19. E. Johnston; S. Bullock; J. Uilk; P. Gatenholm; K. J. Wynne. *Macromolecules* **1999**, *32*, 8173-8182.
20. K. J. Wynne; T. Ho; R. A. Nissan; X. Chen; J. A. Gardella, Jr. *ACS Symposium Series* **1994**, *572*, 64-80.
21. X. Chen; J. A. Gardella, Jr.; T. Ho; K. J. Wynne. *Macromolecules* **1995**, *28*, 1635-1642.
22. J. K. Pike; T. Ho; K. J. Wynne. *Chemistry of Materials* **1996**, *8*, 856-860.
23. S. Bullock; E. E. Johnston; T. Willson; P. Gatenholm; K. J. Wynne. *Journal of Colloid and Interface Science* **1999**, *210*, 18-36.
24. Y. Tezuka; H. Kazama; K. Imai. *Journal of the Chemical Society, Faraday Transactions* **1991**, *87*, 147-152.
25. Y. Tezuka; T. Ono; K. Imai. *Journal of Colloid and Interface Science* **1990**, *136*, 408-414.
26. J. N. Cawse, Ed. *Experimental Design for Combinatorial and High Throughput Materials Development*; Wiley-Interscience: New York, NY, 2003.
27. R. Potyrailo; E. J. Amis, Eds. *High-Throughput Analysis: A Tool for Combinatorial Materials Science*; Kluwer Academic/Plenum Publishers: New York, 2004.
28. D. C. Webster; J. Bennett; S. Kuebler; M. B. Kossuth; S. Jonasdottir. *JCT CoatingsTech* **2004**, *1*, 34-39.
29. J. J. Hoffman; C. M. Leir. *Polymer International* **1991**, *24*, 131-138.
30. P. Majumdar; D. A. Christianson; D. C. Webster. *Polymeric Materials Science and Engineering* **2004**, *90*, 799-800.

Stimuli-Responsive Coatings

Chapter 6

Smart Responsive Coatings from Mixed Polymer Brushes

Sergiy Minko

Chemistry Department, Clarkson University, 8 Clarkson Avenue, Potsdam, NY 13699–5810

Mixed polymer brushes represent a new type of responsive thin polymer films that can switch surface properties upon external stimuli. Mixed brushes are fabricated by the grafting of two unlike polymers to the surface of the same substrate. The switching mechanism is caused by phase segregation of the unlike polymers. In a nonselective environment the polymers segregate laterally and chemical composition of the top layer is affected by the fractions of the polymers in the mixed brush. However, in a selective environment the brush segregates into layered phases where one of the polymers preferentially occupies the top of the brush. This mechanism results in large changes of the top layer chemical composition. The latter affects many important properties of the coatings: wetting behavior, adhesion, adsorption, optical properties, etc.

Introduction

Traditional functions of coatings that encounter the protection from an aggressive environment and contaminations, optical properties, electroconductive properties, improvement of surface mechanical resistance, regulation of friction, biocompatibility, catalytic activity, or anti-microbial functions, etc. recently have been extended to "smart" responsive behavior. A smart coating is tailored in such a way that some of above listed functions can be "switched on" or "switched off" upon change of outside conditions or upon an external signal. The switching upon outside conditions is assigned as a self-adaptive behavior. The switching upon a directed external signal is considered as an "active response". Both responsive phenomena bring valuable functions of the smart coatings. For example, the same material can be hydrophobic or hydrophilic, sticky or non-adhesive, can repeal or attract proteins and cells, can be positively or negatively charged, reflective and non-reflective, conductive or non-conductive, etc. All these transformations can be performed just upon external stimuli. Different external stimuli are used to initiate switching: temperature, humidity, vapors or solutions of organic chemicals, pH and ionic strength in an aqueous environment, electrical or magnetic fields, as well as electromagnetic irradiation (light, for example).

The research area of smart coatings is a rapidly developing field. Major contributions to the field were done exploring polymers. The nature of chain

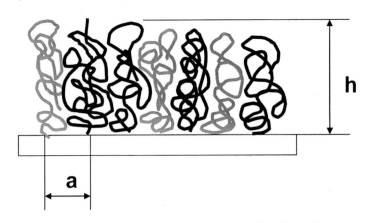

Figure 1. Mixed "planar" polymer brush from "black" and "gray" chains:
h – is the brush thickness, a – is the distance between grafting points.

molecules where conformation is very sensitive to their environment makes them the most suitable materials for responsive coatings. Due to a small contribution of the entropic term into the free energy of mixing even small changes of interactions in polymer solutions or blends can affect phase transitions and, consequently, a strong alteration of material properties. Various aspects of this mechanism have been used to design smart coatings (*1*).

The paper reviews recent results on smart coatings fabricated from grafted polymer layers which were named "mixed polymer brushes" (*2*).

The mixed polymer brush is a thin film (monolayer) constituted of randomly tethered to the same substrate two or more different polymers. The larger difference between properties of these two unlike polymers the larger switching effect can be obtained. The mixed brush is tailored in such a way that external signals affect phase segregation of unlike polymers in the film. A microstructure of the film is switched due to the phase segregation when one polymer is reversibly exposed to the top of the brush, while the second polymer segregated to the grafting surface. Thus, the surface properties of the coatings are affected by the polymer exposed to the top layer. Changes of the mixed brush environment cause changes in the phase segregation mechanism and, therefore, the concentration profile of the polymers in the thin film. All the transitions are reversible unless the film is damaged.

Structure of mixed brushes

Various structural combinations can be designed if two or more different polymers are end tethered (grafted) to the same solid substrate. Several characteristic structures with specific behavior could be identified: planar and spherical mixed brushes, symmetric and asymmetric mixed brushes, random and Y-shaped mixed brushes, block-copolymer brushes.

Mixed brushes are characterized by grafting density (number of grafted polymer chains per surface unit), height, molecular weight of grafted chains, composition (fractions of different grafted polymers), molecular weight distribution of grafted chains, and distribution of distances between grafted points of the polymers (Figure 1). Similar to homobrushes the mixed brush regime is approached if the distance between grafted points is smaller than the gyration radius of the grafted chains. In these conditions the volume excluded effect provokes chains to stretch away from the grafting surface (substrate).

If the radius of curvature of the substrate is of the same order of magnitude (or less) as the brush height the mixed brush is called as a spherical mixed brush. Its behavior is somewhat different from planar brushes (Figure 2a) (*3, 4*) because

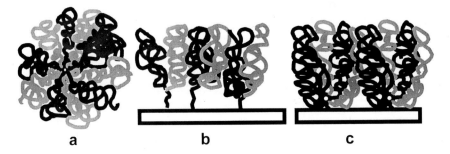

Figure 2. Structure of mixed brushes: spherical brush -a, Y-shape brush –b, grafted spherical mixed brush –c.

in the spherical brush the grafting density decreases with distance from the grafting surface.

Different polymer chains could be randomly grafted to the substrate, or two different polymer arms can emanate from the one grafting point (or from the same spacer). In the latter case the mixed brush is called as a Y-shape brush. (Figure 2b) (5). If a spherical mixed brush (Figure 2a) is grafted to the solid substrate via one arm (or several arms) it can be considered a mixed brush also (Figure 2c). Grafting points for different polymers in the mixed brush may form alternating ordered arrays. In this case the brush could be considered a mixed brush at a macroscopic scale. However, at a microscopic scale this brush consists of ordered homobrush domains of one of the polymers separated by homobrush domains of the second polymer (fabricated using soft lithography Figure 3) (6). Finally, mixed brushes can be patterned (7).

Gradient mixed brushes are represented by grafted films with a monotonous change of the mixed brush composition across the sample (8, 9).

A specific case of mixed brushes is a hybrid brush where the brush layer contains nanoparticles (metallic or oxides) (10, 11). The nanoparticles attached to the brush will modify properties of the coatings as well as they may diffuse in the brush layer upon switching.

Synthesis of mixed brushes

Two major approaches were developed for the synthesis of the mixed brushes. The "grafting from" approach includes (i) grafting of an initiator to the solid substrate, and (ii) polymerization initiated by the grafted initiator. The process is repeated two times to graft two different polymers step by step. The same or two different initiators can be used to graft two different polymers (12-14). The most explored mechanism is the radical polymerization (15-17).

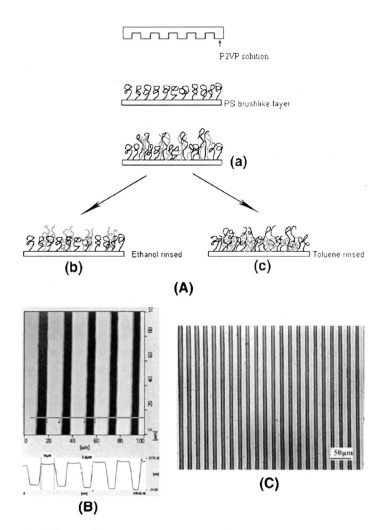

Figure 3. Scheme of microcontact printing of P2VP chains (gray color) on the Si wafer grafted with PS (black color) (A) and the conformation of the brush (a) after rinsing with ethanol (b) and toluene (c), respectively. (B) Topographical and cross-sectional AFM images of the PDMS mold with the stripe dimensions of 10 μm/8.5 μm and (C) Optical micrograph of the P2VP–COOH stripes printed on the Si wafer grafted with PS. (Reproduced from reference 6., Copyright 2004, with permission from Elsevier).

The "grafting to" approach explores end functionalized polymers. The surface of the substrate is chemically modified to introduce complimentary functional groups (usually with silane or thiol chemistry). Then brushes are synthesized via reactions of the complimentary functional groups on the surface with the end-functionalized polymer chains. In this approach the two step grafting process for each end-functionalized polymer is used to synthesize mixed brushes (18).

The composition of the mixed brushes is measured with FTIR ATR and XPS methods (18). Ellipsometric measurements (which follow each grafting step) allow for the quantitative information of each grafting step. Switching behavior of the mixed brushes was studied using in situ experiments where the brush was exposed to solvent and the brush characteristic were recorded employing, for example, ellipsometric, AFM, and Z-potential measurements (19). However, many methods are not suitable for measurements in liquids and vapors. In this case the brush morphology should be rapidly frozen by cooling to the temperature below the glass transition temperature of the grafted polymers to perform "off line" study of the mixed brush (1).

Mixed brush behavior

The mixed brush morphology is balanced by polymer-polymer and polymer-solvent interactions in the grafted layer (20, 21). In non-selective solvents the mixed brush microscopically segregates into lamellar type domains (lateral segregation of the polymer into alternating elongated structural domains) (Figure 4). No a long range order was observed for the lamellar phases of mixed brushes in contrast to block-copolymer systems. Grafting imperfections (various defects, deviations from random grafting) is considered to be a reason for that. Due to this morphology both the polymers are exposed to the top layer of the mixed brush. The surface properties of the film are affected by the combination of properties of both the polymers exposed to the top.

A change of solvent selectivity results in a strong alteration of the morphology and brush properties. In a selective solvent the unfavorable polymer collapses into the clusters. The clusters are surrounded by the swollen favorable polymer. The clusters segregate preferentially to the grafting surface. Thus, the top layer is preferentially occupied by the favorable polymer. The favorable polymer has the major effect on the surface properties of the film. If solvent selectivity is switched (via a change of solvent composition or via a change of temperature) the morphology is reversibly changed. This simple mechanism allows for the switching of the surface properties upon a change of a solvent selectivity. The balance of the interactions in the system can be regulated by the mixed brush composition, molecular weight, or grafting density. Electrostatic interactions are used to regulate interactions in the mixed brushes in an aqueous environment (*19*).

Examples of switching behavior

Switching of wetting behavior

Mixed brushes fabricated from the mixture of hydrophilic and hydrophobic grafted polymers can switch wetting behavior of the film from wetting to nonwetting (*12,18,22*). For example, the brush prepared from polystyrene (PS) and poly (2-vinylpyridine) chains is hydrophilic upon treatment with acidic aqueous solutions (*22*). This brush can be switched into hydrophobic state upon treatment with toluene. If the brush is exposed to toluene, which is a selective solvent for PS, the PS chains preferentially occupy the top of the brush. This morphology of the brush was frozen by a rapid evaporation of toluene. The transition between both the morphologies depends on solvent nature. Many intermediate states of wetting behavior were obtained upon treatment with various solvents (Table 1).

The switching of wetting behavior can be substantially amplified if the mixed brush is grafted onto a rough substrate. For example, the mixed brush prepared from the copolymer of styrene and fluorinated acrylate (PSF) chains and P2VP chains grafted onto the surface of the substrate with a needle-like morphology was shown to be capable for switching from a very hydrophilic state in acidic water (complete sheeting) to an ultrahydrophobic state (low hysteresis, water contact angle 160^0) upon treatment with toluene (*23*). The microscopic needles introduce a surface roughness where a Cassie wetting regime is approached (*24*). The drop of water deposited on the top of the material is in contact mainly with trapped air.

Switching of adhesion

Switching the surface composition of the films may strongly affect adhesion. For example, PSF forms a nonsticky thin film. The mixed brush constituted from PSF and P2VP is non sticky upon treatment with toluene. If the brush is exposed to acidic aqueous solution the adhesive pyridine functional groups are exposed to the top. Mechanical tests show that adhesive forces between the brush surface and the adhesive tape change in two folds upon switching the brush (Figure 5) (*23*).

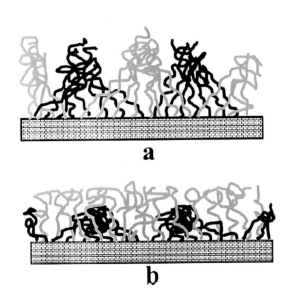

Figure 4. Schematic illustration of two possible morphologies of mixed brush irreversibly grafted to solid substrates (cross section of the layer): lamellar morphology in a nonselective solvent (a); cluster morphology in a solvent poor for the black chains (b) (Reproduced from reference 21. Copyright 2003 American Chemical Society).

Table 1. Advanced contact angles (CA) of water and root mean square roughness (RMS) on the surface of binary mixed brushes PS-P2VP (B1 -56%PS, B2- 71%PS) after exposure to different solvents

Solvent	B1			B2		
	CA, deg	Fraction of PS on the top of the layer	RMS, nm	CA, deg	Fraction of PS on the top of the layer	RMS, nm
toluene	85	0.79	1.9	90	1.0	2.8
chloroform	81	0.62	0.5	77	0.45	0.7
ethanol	70	0.16	1.4	68	0.08	2.0
Water pH7	74	0.32		74	0.32	
Water pH 2.5	48			36		

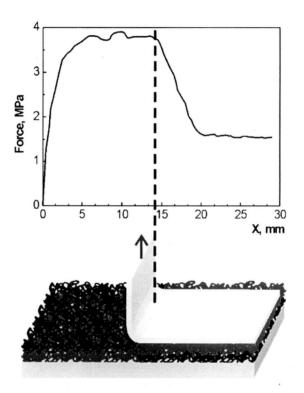

Figure 5. Switching adhesion with PSF-P2VP mixed brush: the plot presents the change of the force applied to the Tesa band defoliated from the brush vs distance (X) from the starting point. The dashed line marks the border between ultrahydrophobic (orange) and hydrophilic (blue) areas on the brush (Reproduced from reference 23. Copyright 2003 American Chemical Society).

Switching of adsorption.

The switching of adsorption is relevant to the switching of wetting behavior and adhesion. All these properties depend on the surface chemical composition and morphology of the mixed brush. The mixed polyelectrolyte brush represents a very nice example of the substrate which explores a range of various surface forces tuned upon outside signals: van der Waals, hydrophobic interactions, hydrogen bonds, and electrostatic interactions. One of the most interesting applications of the responsive coatings is tuning protein adsorption. Figure 6 illustrates an example of complex interactions between the mixed polyelectrolyte brush (fabricated from P2VP and PS, and from P2VP and polyacrylic acid (PAA)) and mioglobin. Adsorption was studied from aqueous solutions at different pH. In the pH range between 5 and 7 (around isoelectric point (IEP)) the mixed brushes are weakly charged. In this case hydrophobic interactions dominate and affect the strong adsorption of the protein. At lower and higher pH values the electrostatic repulsion strongly modifies adsorption, because the brush and the protein are oppositely charged. In this case the balance between interfacial forces was switched with pH signal. The adsorption value was dramatically changed upon switching.

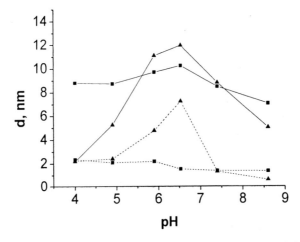

Figure 6. Adsorption of mioglobin (IEP 7.2) on the P2VP-PS (IEP 5.9) (squares) and P2VP-PAA (IEP 4.9) (triangles) mixed brushes presented as protein layer thickness vs. .pH ((0.01 M PBS buffer -solid lines, 0.001 M PBS-dashed lines).

Switching of optical properties

Morphological changes in the responsive coatings can be explored for switching optical properties. Switching between different mechanisms of phase segregation in mixed brushes affects changes in refractive index and coefficients of reflectivity of the thin film. For example, the refractive index of the mixed P2VP-PAA polyelectrolyte brush was tuned by a change of pH in an aqueous environment (Figure 7) (*19*). Switching between segregated phases may change the roughness of the thin film. This change influences spectra of the reflected light which appear as a change of the thin film color (*25*).

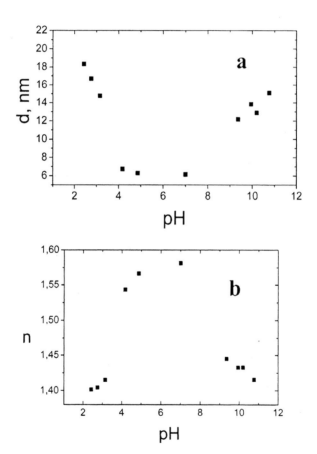

Figure 7. Influence of pH on the (a) thickness (d) and (b) refractive index (n) of the swollen PAA-P2VP brush (Reproduced from reference 19. Copyright 2003 American Chemical Society).

Applications in devices

A change of physical properties of thin polymer films caused by transitions between different phase segregation mechanisms can be explored for the fabrication of smart surfaces and for using them to design various devices. For example, the walls of microchannels in a microfluidic device coated with the mixed brush can act as a microscopic valve. Temperature, solvent selectivity, or pH were used as external signals to switch between hydrophilic and hydrophobic states of the channels (Figure 8) (7). Thus, in a hydrophobic state channels were closed for aqueous solutions blocking the liquid flow, while in a hydrophilic state of the channels the liquid flowed through the device.

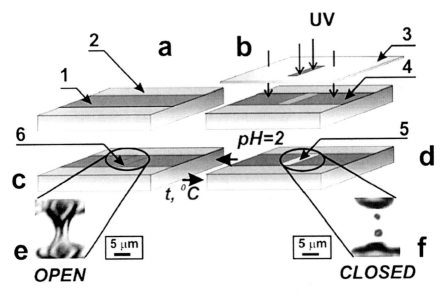

Figure 8. Fabrication of a switchable channel. The images e and f show open and closed states of the switchable channel, respectively, as they appear in optical microscopy (Reproduced from reference 7. Copyright 2003 American Chemical Society).

Combination of the polymer brush with metal nanoparticles (hybrid brush) allowed for the fabrication of nanosensors. The responsive behavior of the brush and switching of the optical properties of the layer was used to enhance the band shift in the transmission surface plasmon spectra of the brush-nanoparticle composite (Figure 9) (26).

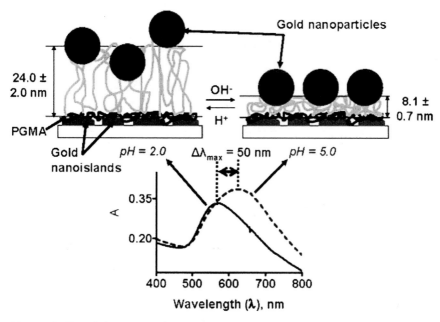

Figure 9. Top: Schematics of the reversible pH change-induced swelling of gold nanoparticle-coated P2VP brushes. Bottom: T-SPR spectra of gold nanoislands (containing P2VP polymer brush and gold nanoparticles) at pH 2.0 and 5.0 (Reproduced from reference 26. Copyright 2004 American Chemical Society).

Conclusions

The responsive polymer coatings fabricated from mixed polymer brushes show remarkable behavior and diverse potential applications. Further investigations will be performed towards the design of mixed brushes to understand and regulate interactions in complex biological systems. Many promising applications are expected from the design of smart coatings for cloth, textiles, paper, and glass. The obtained knowledge creates the basis for the development of new approaches for microfluidic technologies and sensors.

Acknowledgement

I am grateful for the contributions of my colleagues and students: Manfred Stamm, Marcus Mueller, Constantinos Tsitsilianis, Igor Luzinov, Vladimir

Tsukruk, Andreas Scholl, Alexander Sidorenko, Janos Fendler, Eliza Hutter, Iryna Tokatreva, Valeriy Luchnikov, Denis Usov, Nikolai Houbenov, Leonid Ionov, Mikhail Motornov, Igor Tokarev, Bogdan Zdyrko, Satish Patil, and many others who help us to perform our research at the Institute for Polymer Research Dresden, and Clarkson University.

References

1. Luzinov, I.; Minko, S.; Tsukruk, V. V. *Prog. Polym. Sci.* **2004**, *29*, 635-698.
2. Minko, S.; Müller, M.; Luchnikov, V.; Motornov, M.; Usov, D.; Ionov, L.; Stamm, M. In *Polymer Brushes*; Advincula, R. C.; Brittain, W. J.; Caster, K. C.; Rühe, J., Eds.; Wiley-VCH: Weinheim, 2004; pp 403-425.
3. Gorodyska, G.; Kiriy, A.; Minko, S.; Tsitsilianis, C.; Stamm, M. *Nano Lett.* **2003**, *3*, 365-368.
4. Kiriy, A.; Gorodyska, G.; Minko, S.; Stamm, M.; Tsitsilianis, C. *Macromolecules* **2003**, *36*, 8704-8711.
5. Julthongpiput, D.; Lin, Y. H.; Teng, J.; Zubarev, E. R.; Tsukruk, V. V. *Langmuir* **2003**, *19*, 7832-7836.
6. Yu, K.; Cong, Y.; Fu, J.; Xing, R. B.; Zhao, N.; Han, Y. C. *Surf. Sci.* **2004**, *572*, 490-496.
7. Ionov, L.; Minko, S.; Stamm, M.; Gohy, J. F.; Jerome, R.; Scholl, A. *J. Am. Chem. Soc.* **2003**, *125*, 8302-8306.
8. Ionov, L.; Sidorenko, A.; Stamm, M.; Minko, S.; Zdyrko, B.; Klep, V.; Luzinov, I. *Macromolecules* **2004**, *37*, 7421-7423.
9. Ionov, L.; Houbenov, N.; Sidorenko, A.; Stamm, M.; Luzinov, I.; Minko, S. *Langmuir* **2004**, *20*, 9916-9919.
10. Santer, S.; Rühe, J. *Polymer* **2004**, *45*, 8279-8297.
11. Boyes, S.; Akgun, B.; Brittain, W. J.; Foster, M. D. *Abstr. Am. Chem. Soc.* **2004**, *227*, U520-U520.
12. Sidorenko, A.; Minko, S.; Schenk-Meuser, K.; Duschner, H.; Stamm, M. *Langmuir* **1999**, *15*, 8349-8355.
13. Zhao, B. *Polymer* **2003**, *44*, 4079-4083.
14. Zhao, B.; He, T. *Macromolecules* **2003**, *36*, 8599-8602.
15. Minko, S.; Gafijchuk, G.; Sidorenko, A.; Voronov, S. *Macromolecules* **1999**, *32*, 4525-4531.
16. Minko, S.; Sidorenko, A.; Stamm, M.; Gafijchuk, G.; Senkovsky, V.; Voronov, S. *Macromolecules* **1999**, *32*, 4532-4538.
17. Sidorenko, A.; Minko, S.; Gafijchuk, G.; Voronov, S. *Macromolecules* **1999**, *32*, 4539-4543.
18. Minko, S.; Patil, S.; Datsyuk, V.; Simon, F.; Eichhorn, K. J.; Motornov, M.; Usov, D.; Tokarev, I.; Stamm, M. *Langmuir* **2002**, *18*, 289-296.

19. Houbenov, N.; Minko, S.; Stamm, M. *Macromolecules* **2003**, *36*, 5897-5901.
20. Minko, S.; Muller, M.; Usov, D.; Scholl, A.; Froeck, C.; Stamm, M. *Phys. Rev. Lett.* **2002**, *88*, 035502.
21. Minko, S.; Luzinov, I.; Luchnikov, V.; Muller, M.; Patil, S.; Stamm, M. *Macromolecules* **2003**, *36*, 7268-7279.
22. Minko, S.; Usov, D.; Goreshnik, E.; Stamm, M. *Macromol. Rapid Comm.* **2001**, *22*, 206-211.
23. Minko, S.; Muller, M.; Motornov, M.; Nitschke, M.; Grundke, K.; Stamm, M. *J. Am. Chem. Soc.* **2003**, *125*, 3896-3900.
24. Bico, J.; Tordeux, C.; Quere, D. *Europhys. Lett.* **2001**, *55*, 214–220.
25. Lemieux, M.; Usov, D.; Minko, S.; Stamm, M.; Shulha, H.; Tsukruk, V. V. *Macromolecules* **2003**, *36*, 7244-7255.
26. Tokareva, I.; Minko, S.; Fendler, J. H.; Hutter, E. *J. Am. Chem. Soc.* **2004**, *126*, 15950-15951.

Chapter 7

Surface-Catalyzed Growth of Polymer Films That Respond to pH

Dongshun Bai, Brian M. Habersberger, and G. Kane Jennings*

Department of Chemical Engineering, Vanderbilt University,
Nashville, TN 37235

We have developed a new surface-catalyzed method to grow ester-functionalized polymethylene (PM) films from gold surfaces by exposure to a mixture of diazomethane (DM) and ethyl diazoacetate (EDA) in ether at 0 °C. The ester groups are randomly distributed within the copolymer at a concentration that can be tuned by varying the ratio of EDA to DM in solution. The ester side chains of these films can be modified by exposure to base to produce carboxylic acid groups that become deprotonated under basic conditions to function as a pH-responsive gate. At low pH, the polymer film is predominately hydrophobic due to the high content of PM and exhibits a low capacitance. Upon stepping pH from 4 to 11, the acid groups become deprotonated and charged, greatly enhancing the permeability of water and boosting the capacitance by a factor of ~400.

Introduction

Polymer films that respond to pH by altering structure, barrier properties, and/or surface properties have high potential to impact chemical sensors,[1-3] membrane separations,[4-7] drug delivery,[8] and smart, dynamic surfaces.[9] These smart materials have been prepared in many forms, including self-assembled monolayers,[5,6] Langmuir-Blodgett (LB) films,[9] and cast polymer films.[4,8] The polymeric materials are generally more interesting than the monolayer counterparts due to the capability for a larger response to a pH stimulus. Specific examples of pH-responsive polymer materials include hydrogels that swell upon changes in pH to facilitate drug transport[8] and carefully tailored copolymer membranes that are sensitive to a pH stimulus to enable ions of specific charge or molecules of specific properties to pass.[4] To date, responsive polymers have been prepared predominately by solution-phase synthesis to yield well-defined bulk materials that can be diluted in solvent and cast onto surfaces to prepare films. While appropriate for applications where thick films are required on 2-D surfaces, this casting approach is less effective for creating patterned films, controlling film thickness with nanometer-level precision, or coating irregular surfaces.

Surface-initiated polymerizations[10] offer many advantages over traditional coating methods for the preparation of responsive films. Some of these benefits include (1) the ability to prepare uniform, conformal coatings on objects of any shape,[11] (2) precise control over film thickness,[12] from a few nanometers up to the micron level,[11,13] (3) tunable grafting densities, based on the surface coverage of the initiator, and (4) control over depth-dependent composition by growing additional blocks to prepare copolymer films in which one block or region of the film is responsive.[14-16] For example, Brittain and coworkers[14,17] have prepared di- and triblock copolymer brushes that change morphology and surface composition upon exposure to a good solvent for the inner block(s). In addition, Lopez and coworkers[18,19] have grown thermally responsive poly(N-isopropylacrylamide) films from surfaces to investigate temperature-switchable membranes and the temperature-programmed release of biofilms. Combined, this prior work demonstrates that surface-initiated strategies can be used to grow brush-like films that respond to solvent and temperature.

Herein, we utilize a unique surface-initiated approach known as surface-catalyzed polymerization[13,20,21] to engineer a new class of pH-responsive copolymer films. The films are prepared by exploiting a selective catalysis at gold surfaces. Briefly, we have recently reported that exposure of gold substrates to a dilute solution containing diazomethane (DM) and ethyl diazoacetate (EDA) (Figure 1) results in a catalyzed copolymerization of a film containing linear polymethylene (PM) with randomly distributed ethyl ester side groups (denoted as PM-CO2Et).[22] The film propagation occurs from the metal-polymer interface and the rate of film growth is constant with time, suggestive of a controlled polymerization. These ester side groups can be hydrolyzed to carboxylic acids (PM-CO2H) or converted to other ionizable groups to provide pH responsive behavior. In fact, we have demonstrated pH-dependent wettability for PM-CO2H films.[22]

The proposed strategy toward pH-responsive films has several advantages over other approaches. (1) Surface-catalyzed polymerizations offer the ability to rapidly grow films with controlled thicknesses; in fact, film thickness increases linearly with time.[22] (2) The film growth is selective in that polymerization occurs on gold but not on most other materials (silicon, silver, aluminum, plastics, etc.), enabling straightforward patterning of films by directed growth.[13] (3) An initiator is not required as the film growth propagates directly from the metal surface. (4) Film composition can easily be tuned to affect the film response without requiring a completely new macromolecular synthesis. (5) The pH response can be assessed by either capacitance- or piezoelectric-based methods. Using this surface-catalyzed approach, a careful control over film composition and thickness can be employed to tailor the onset, magnitude, and rate of the response in barrier properties. In the current manuscript, we are especially interested in the effect of pH on the capacitive properties of the films, since capacitance changes can form the basis for a successful sensor. We have designed PM-CO2H films to consist predominately (>95%) of polymethylene (PM) so that the film is hydrophobic in the uncharged state. When the acid groups become deprotonated or charged, the water solubility of the functional groups should increase by ~4 orders of magnitude[23] to greatly alter the film capacitance.

Experimental

Materials

Potassium hydroxide, Diazald (N-methyl-N-nitroso-P-toluenesulfonamide), ethyl diazoacetate (EDA), and poly(ethylene-co-ethylacrylate) (PEEA) were used as received from Aldrich (Milwaukee, WI). Ethyl ether was obtained from EMD Chemicals (Gibbstown, NJ). Potassium hydrogen phthalate, sodium bicarbonate, and benzoic acid were used received from Fisher (Fair Lawn, NJ). Sodium hydroxide and hydrochloric acid were used as received from EM Science (Gibbstown, NJ). Gold shot (99.99%) and chromium-coated tungsten filaments were obtained from J&J Materials (Neptune City, NJ) and R.D. Mathis

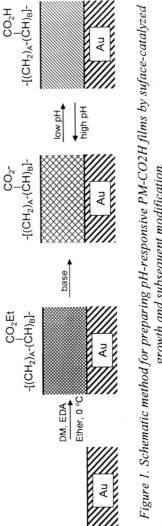

Figure 1. Schematic method for preparing pH-responsive PM-CO2H films by suface-catalyzed growth and subsequent modification.

(Signal Hill, CA), respectively. Silicon (100) wafers (Montco Silicon; Spring City, PA) were rinsed with ethanol and deionized water and dried with nitrogen. Ethanol (absolute) was used as received from AAPER (Shelbyville, KY). Nitrogen gas was obtained from J&M Cylinder Gas, Inc (Decatur, AL). Deionized water (16.7 MΩ·cm) was purified with a Modu-Pure system (Continental Water Systems Corporation; San Antonio, TX) and used for rinsing.

Preparation of Gold Substrates

Gold substrates were prepared by evaporating chromium (100 Å) and gold (1250 Å) in sequence onto silicon (100) wafers at rates of 1-2 Å s^{-1} in a diffusion-pumped chamber with a base pressure of 3 x 10^{-6} torr. After removal from the evaporation chamber, the wafers were typically cut into 1 cm x 4 cm pieces, rinsed with ethanol, and dried in a stream of N$_2$ gas.

Preparation of Diazomethane

DM was carefully prepared according to a literature procedure[24] and diluted with ether at 0 °C to prepare solutions of different concentration. CAUTION: Diazomethane is toxic and potentially explosive and should be handled carefully![24] The concentration of DM was determined by titration with benzoic acid.

Preparation of Polymer Films

Polymer films were formed by exposure of gold-coated silicon substrates (ca. 4 cm x 1 cm) to ether solutions containing 1 mM DM and 40 mM EDA at 0 °C for 24 h. Film growth was carried out in capped 20 mL vials and only one substrate was placed in each vial. Upon removal, the samples were rinsed with ether and dried in a stream of nitrogen.

Hydrolysis

Hydrolysis of the copolymer films was carried out in a solution of 0.5 M KOH in ethanol at 60 °C for desired times. The hydrolyzed samples were rinsed with ethanol and DI water and dried in a N$_2$ stream.

Characterization Methods

Polymer film properties were evaluated using the following methods. Reflectance absorption infrared spectroscopy (RAIRS) was performed using a Bio-Rad Excalibur FTS-3000 infrared spectrometer. The p-polarized light was incident at 80° from the surface normal. The instrument was run in single reflection mode and equipped with a Universal sampling accessory. A liquid nitrogen-cooled, narrow-band MCT detector was used to detect reflected light. Spectral resolution was 2 cm^{-1} after triangular apodization. Each spectrum was accumulated over 1000 scans with a deuterated octadecanethiol-d_{37} self-assembled monolayer on gold as the background.

The molar ester contents of the surface-catalyzed copolymer films were determined from reflectance IR spectra using the integrated area ratios for the carbonyl stretching peak at 1735 cm^{-1} and the combined methylene stretching peaks (symmetric and asymmetric) at 2850 and 2920 cm^{-1}, respectively. The basis for the calculation of ester content was the peak area ratio obtained for a cast film of a commercially available random copolymer (poly(ethylene-co-ethyl acrylate) (PEEA)). The PEEA has a known 18 wt% ethyl acrylate content (2.9% (molar) ethyl ester; 97.1% -CH$_2$-) and exhibited a C=O:CH$_2$ peak area ratio of 0.26. This analysis assumes that the peak area ratio scales linearly with the molar ester content within the film.

Ellipsometry measurements were obtained on a J.A. Woollam Co. M-2000DI variable angle spectroscopic ellipsometer with WVASE32 software for modeling. Measurements at three spots per sample were taken with light incident at a 75° angle from the surface normal using wavelengths from 250 to 1000 nm. Optical constants for a bare gold substrate, cut from the same wafer as the samples to be characterized, were measured by ellipsometry and used as the baseline for all polymer film samples. Film thickness of the polymer layer on samples was determined using a Cauchy layer model. Since the copolymer films are PM rich (ester content < 5%), we set the refractive index for the film to 1.5, consistent with the ranges measured for polyethylene.[25]

Electrochemical impedance spectroscopy (EIS) was performed with a Gamry Instruments CMS300 impedance system interfaced to a personal computer. A flat-cell (EG&G Instruments) was used to expose only 1 cm^2 of each sample to an aqueous solution containing electrolyte and redox probes while preventing sample edges from being exposed. The electrochemical cell consisted of a buffered aqueous solution (0.1 M ionic strength) at pH 4 or pH 11 with a Ag/AgCl/saturated KCl reference electrode, a gold substrate counter electrode, and a gold substrate containing the film to be studied as the working electrode. All data were collected in the range from 10^{-1} to 10^4 Hz using 10 points per decade and were fit with an appropriate equivalent circuit model (vide infra) to determine resistance and capacitance values.

Results and Discussion

Upon exposure of gold substrates to a solution containing 1 mM DM and 40 mM EDA in ether at 0 °C for 24 h, we obtained a 320 nm polymer film that yields a reflectance IR spectrum as shown in Figure 2a. The film exhibits asymmetric and symmetric methylene stretching peaks at 2920 and 2851 cm^{-1}, respectively, consistent with polycrystalline packing of the polymer chains within the film. Figure 2a also reveals important compositional information with signatures for methyl (-CH$_3$) stretching at ~2979 cm^{-1}, methyl bending at 1375 cm^{-1}, and carbonyl stretching at 1736 cm^{-1}. These peaks are consistent with a structure as shown in Figure 1 where ethyl ester groups are attached to a predominately PM chain. From the intensity of the ester peak, in comparison with a cast film of PEEA with known composition, we estimate that the molar ester content is 3.5% for this film. This percentage can be adjusted by altering the ratio of DM to EDA in solution.[22]

Upon exposure of a 320-nm PM-CO$_2$Et film (3.5 % molar ester content) to a 0.5 M solution of KOH in ethanol at 60 °C for 36 h, the ester side groups within the film are mostly (80 %) converted to carboxylate groups as evidenced by the appearance of a peak at 1569 cm^{-1}, characteristic of a carboxylate carbonyl, and the dramatic decrease in the ester carbonyl peak in the IR spectrum of Figure 2c. The magnitude of the reduction in the ester peak intensity provides an estimate of the conversion in this hydrolysis reaction. Figure 3 shows the conversion as a function of time upon exposure of a 320 nm PM-CO2Et film to 0.5 M KOH in ethanol at 60 °C. Over 75% of ethyl esters are converted to carboxylates during the first 24 h, and very gradual increases occur during the next 24 h. Importantly, the film is stable to this hydrolysis based on the constant intensities of methylene stretching peaks in the IR spectra for films hydrolyzed for 36 h (Figure 2). Exposure times beyond 36 h resulted in minimal enhancement of conversion but also seemed to degrade the films based on visual inspection and changes in the IR spectra and electrochemical impedance spectra. We are currently investigating alternative solvents to enable higher conversions while eliminating film degradation.

Figures 2b and 2c show RAIR spectra for the film after hydrolysis and after exposure to pH 4 or pH 11 buffer solutions. The carboxylate carbonyl at 1569 cm^{-1} (pH 11) shifts to 1710 cm^{-1} after exposure to a pH 4 buffer (Figure 2b). The films can be converted between charged (deprotonated) and uncharged (protonated) states repeatedly and reversibly by exposure to the respective buffer solutions. The hydrolysis and subsequent exposure to buffer solutions does not affect the structure of the film based on the similar peak positions (2921 cm^{-1}) and intensities for CH$_2$ stretching bands. We have also observed that the rate of this transformation is rapid and accomplished within ~10 s for a 320 nm film, based on in situ, time-dependent capacitance measurements. The rapid kinetics suggests that the films would exhibit fast response for applications as sensors or smart films.

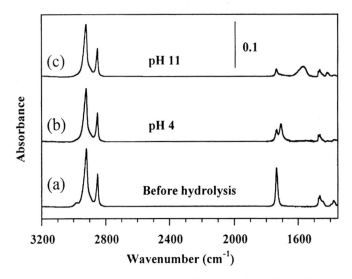

*Figure 2. Reflectance infrared spectra of a PM-CO2Et film before hydrolysis
(a) and after hydrolysis followed by exposure to pH 4 (b) and pH 11 (c) buffer
solutions.*

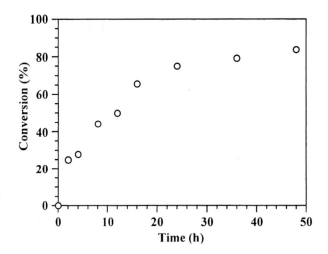

*Figure 3. Percent conversion of PM-CO2Et to PM-CO2H upon exposure to a
0.5 M solution of KOH in ethanol at 60 °C.*

To assess film performance, we have obtained electrochemical impedance spectra for a 200 nm PM-CO2H film (2.9% CO2H) upon exposure to aqueous buffer solutions at pH 4 and pH 11 (Figure 4). At pH 4, the polymer film provides a barrier against water and ion permeation as evidenced by the elevated values for impedance modulus. This spectrum is fit by an equivalent circuit model that we have described previously,[26] which consists of two time constants, one corresponding to the polymer film and a second due to the polymer/metal interface. At pH 4, the impedance due to the polymer film dominates the impedance due to the metal/polymer interface; thus, only a single time constant due to the polymer film is observed in the spectrum, which is fit to yield a value of 3.6×10^{-8} F/cm^2 for film capacitance and 8×10^4 $\Omega \cdot cm^2$ for film resistance. At pH 11, the acid groups within the film are deprotonated to carboxylates. This ionization of the film greatly increases water permeation and results in a dramatic change in the impedance spectrum. The capacitance and resistance due to the polymer film are not visible or measurable components of this spectrum, and the impedance behavior resembles that for an uncoated electrode, also shown in Figure 4. As a result, the spectrum for the film at pH 11 is fit by a simpler one-time-constant Randles model circuit,[26] which contains elements for the capacitance and resistance at the polymer/metal interface and the solution resistance. A fit of the spectrum recorded at pH 11 yields an interfacial capacitance value of 1.5×10^{-5} F/cm^2, which is of the same order as the value of $\sim 4.4 \times 10^{-5}$ F/cm^2 for a bare gold surface, indicating that the polymer film is likely saturated with water at high pH. This interfacial capacitance at pH 11 is a factor of ~400 higher than the capacitance for the polymer film measured at pH 4. The shift from dominant film capacitance (pH 4) to interfacial capacitance (pH 11) and the large magnitude of this pH-induced response are due to the dramatic enhancement in water solubility ($\sim 10^4$) of CO_2^- vs CO_2H groups and the low fractional content of these functional groups within the film, which assures a hydrophobic, water-resistant barrier film in the uncharged state.

To assess the reversibility of the film response, we have measured capacitances of this same film upon repeated sequential exposures to buffer solutions of pH 4 and pH 11, using electrochemical impedance spectroscopy. Figure 5 shows the measured capacitance upon exposure of the films to each buffer solution for 30 min in a cyclic manner. Upon cycling pH, the film returns to the initial capacitance value, indicating the reversible nature of the pH response. The PM-CO2Et film (control, not shown) did not exhibit changes in capacitance when the pH was stepped. These results demonstrate that the PM-CO2H films respond to pH changes and that the films show promise in sensing applications given the large magnitude (high signal to noise) and reversibility of the response in capacitance.

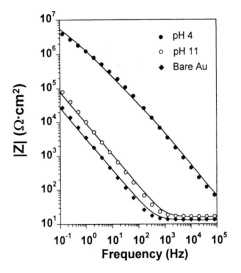

Figure 4. Electrochemical impedance spectra for a 200 nm PM-CO2H film (2.9% CO2H) on gold in aqueous buffer solutions. The spectrum for an uncoated gold substrate is shown for comparison. Curves represent fits of the data using an equivalent circuit model.

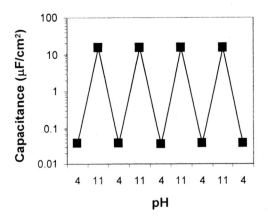

Figure 5. Measured capacitance upon repeated sequential exposures of a PM-CO2H film to pH 4 and pH 11 buffer solutions.

Conclusions

We have prepared pH-responsive carboxyl-modified polymethylene films by a surface-catalyzed polymerization and subsequent hydrolysis. The film is uncharged and hydrophobic at low pH but becomes charged at high pH and enables significant water permeation as evidenced by ~400-fold increases in capacitance. The surface-catalyzed preparation of these films enables selective growth on gold surfaces of any shape, provides a precise control over film thickness, and offers the ability to tune the concentration of the pH-sensitive groups within the film. Future work will explore the effects of incremental pH changes and -CO2H composition on film response.

Acknowledgments

We gratefully acknowledge financial support from the Donors of the American Chemical Society Petroleum Research Fund (ACS-PRF #38553-AC5) for support of our work.

Bibliography

1. Richter, A.; Bund, A.; Keller, M.; Arndt, K.-F., *Sensors and Actuators B* **2004**, 99, 579-585.
2. Lakard, B.; Herlem, G.; Labachelerie, M. d.; Daniau, W.; Martin, G.; Jeannot, J.-C.; Robert, L.; Fahys, B., *Biosensors and Bioelectronics* **2004**, 19, 595-606.
3. Gerlach, G.; Guenther, M.; Suchaneck, G.; Sorber, J.; Arndt, K.-F.; Richter, A., *Macromol. Symp.* **2004**, 210, 403-410.
4. Hester, J. F.; Olugebefola, S. C.; Mayes, A. M., *J. Membrane Science* **2002**, 208, 375-388.
5. Lee, S. B.; Martin, C. R., *Anal. Chem.* **2001**, 73, 768-775.
6. Hou, Z.; Abbott, N. L.; Stroeve, P., *Langmuir* **2000**, 16, 2401-2404.
7. Ito, Y.; Park, Y. S.; Imanishi, Y., *Langmuir* **2000**, 16, 5376-5381.
8. Roy, I.; Gupta, M. N., *Chemistry and Biology* **2003**, 10, 1161-1171.
9. Zhu, X.; DeGraaf, J.; Winnik, F. M.; Leckband, D., *Langmuir* **2004**, 20, 1459-1465.
10. Jennings, G. K.; Brantley, E. L., *Adv. Mater.* **2004**, 16, 1983-1994.
11. Rutenberg, I. M.; Scherman, O. A.; Grubbs, R. H.; Jiang, W.; Garfunkel, E.; Bao, Z., *J. Am. Chem. Soc.* **2004**, 126, 4062-4063.

12. Huang, W.; Kim, J.-B.; Bruening, M. L.; Baker, G. L., *Macromolecules* **2002**, 35, 1175-1179.
13. Guo, W.; Jennings, G. K., *Adv. Mater.* **2003**, 15, 588-591.
14. Boyes, S. G.; Brittain, W. J.; Weng, X.; Cheng, S. Z. D., *Macromolecules* **2002**, 35, 4960-4967.
15. Kim, J.-B.; Huang, W.; Bruening, M. L.; Baker, G. L., *Macromolecules* **2002**, 35, 5410-5416.
16. Matyjaszewski, K.; Miller, P. J.; Shukla, N.; Immaraporn, B.; Gelman, A.; Luokala, B. B.; Siclovan, T. M.; Kickelbick, G.; Vallant, T.; Hoffmann, H.; Pakula, T., *Macromolecules* **1999**, 32, (26), 8716-8724.
17. Granville, A. M.; Boyes, S. G.; Akgun, B.; Foster, M. D.; Brittain, W. J., *Macromolecules* **2004**, 37, 2790-2796.
18. Fu, Q.; Rao, G. V. R.; Ista, L. K.; Wu, Y.; Andrzejewski, B. P.; Sklar, L. A.; Ward, T. L.; Lopez, G. P., *Advanced Materials* **2003**, 15, (15), 1262.
19. Ista, L. K.; Perez-Luna, V. H.; Lopez, G. P., *Applied and Environmental Microbiology* **1999**, 65, 1603-1609.
20. Seshadri, K.; Atre, S. V.; Tao, Y.-T.; Lee, M.-T.; Allara, D. L., *J. Am. Chem. Soc.* **1997**, 119, 4698-4711.
21. Guo, W.; Jennings, G. K., *Langmuir* **2002**, 18, 3123-3126.
22. Bai, D.; Jennings, G. K., *J. Am. Chem. Soc.* **2005**, 127, 3048-3056.
23. Schwarzenbach, R.; Gschwend, P.; Imboden, D., *Environmental Organic Chemistry*. ed.; Wiley: New York, 1993.
24. Aldrich Technical Information Bulletin Number AL-180, 1993.
25. Seferis, J. C., "Refractive Indices of Polymers" In *Polymer Handbook;* Brandrup, J.; Immergut, E. H.; Grulke, E. A., Eds.; Wiley: New York, 1998.
26. Bai, D.; Habersberger, B. M.; Jennings, G. K., *J. Am. Chem. Soc.* **2005**, 127, 16486-16493.

Chapter 8

Pressure Sensing Paints Based on Fluoroacrylic Polymers Doped with Phosphorescent Divalent Osmium Complexes

Brenden Carlson and Gregory D. Phelan

Chemistry Department, University of Washington, Box 351700, Seattle, WA 98195

Pressure sensitive paints (PSP) have proved to be revolutionary in the design of aircraft, cars, trucks, and other vehicles. In this study we incorporate highly phosphorescent divalent osmium complexes into PSP. The divalent osmium complexes were of the form $[Os(N-N)_2 L-L] (PF_6^-)_2$ where N-N was a derivative of 1,10-phenanthroline, and L-L was either cis-1,2-bis(diphenylphosphino)ethene, cis-1,2-vinylenebis-(diphenylarsine), or bis(diphenylarseno) methane. The complexes were dissolved into poly(2-[ethyl-[(heptadecafluorooctyl)-sulfonyl]amino]ethylmethacrylate) at a concentration of 1 mg complex to 1000 mg of polymer. The paints were tested for pressure sensitivity, temperature dependence, and photo-degradation. The paints featured strong pressure response, and luminescence intensity change by up to a factor of two. The temperature dependence was measured as low as -0.1% $^{\circ}C^{-1}$. Many of the complexes exhibited almost non-existent photo-degradation.

Introduction

Pressure sensing paints (PSP) comprise of a luminescent dye, usually one that is phosphorescent, dissolved into, or is part of a polymer matrix. The sensing functions by oxygen quenching of the excited state; hence, emission intensity varies with changes in oxygen concentration or pressure. When molecule "A" absorbs quanta of energy (Figure 1), it is promoted into an excited state. From this excited state, there are a number of pathways for the molecule to return to the ground state. All of the pathways have consequences as to how the PSP performs. The processes of luminescence and the bimolecular quenching are pathways that allow a PSP to function. However, non-luminescent deactivation of the excited state and photochemical decomposition of the molecule are two other processes that may occur and may lead to significant error in pressure measurements. Several luminescent metal complexes have been used for PSP. These include complexes of ruthenium,[1] iridium,[2] and platinum.[3] We report divalent osmium complexes as the luminophore in PSP.

There are several excited states for metal complexes. The excited states are the ligand-centered state, metal centered state, and the charge transfer state. The ligand-centered (LC) state may be defined as transitions such as $\pi-\pi^*$ that take place on the ligand. For a complex with an octahedral coordination sphere such as the ones being presented in this discussion, the metal centered (MC) state may be defined as a transition between filled d_{xy}, d_{xz}, d_{yz} ($d\pi$) and the empty d_z^2, $d_{x^2-y^2}$ ($d\sigma^*$). Charge transfer states may be defined as a transfer of charge from a filled metal orbital to an empty orbital on the ligand, or filled ligand orbitals to empty orbitals on the metal. For the complexes being discussed in this presentation, the charge transfer bands may be called metal to ligand charge transfer (MLCT) and defined by a transition between filled $d\pi$ orbitals and empty π^* orbitals on the ligand. Of these states, the MLCT and the LC states may be luminescent; however, the MC state is usually not luminescent.[4] Each of these states and their relative energy levels to each other may have bearing upon the deactivation processes.

The relation of the excited states to each other depends upon the metal and ligands used to form the complexes. The transition energy of the MC state is dependent upon the location of the metal in the periodic table. Typically, the MC state is lowest in energy for the first row transition metals. Either the charge transfer or ligand centered states are lowest in energy for the second and third row metals. Thus, tris-phenanthroline complexes of iron are usually non-luminescent where as the tris-phenanthroline complexes for either ruthenium or osmium may be luminescent. Even though the MC state is higher in energy for osmium and ruthenium complexes, it still may have an effect on the luminescent properties of the complex. The MC states that are energetically higher than

MLCT or LC states may become thermally populated, causing loss of emission with increasing temperature, and possibly photo-degradation. Thus, separation the energy level of the MC state from the energy levels of the emitting states may minimize the population of the MC state; thus, reducing photo-degradation and temperature dependence of the complex.

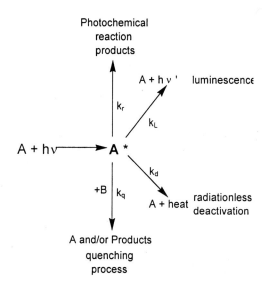

Figure 1. Excitation and some deactivation pathways of an excited molecule.

Experimental

FAB Synthesis

Poly(2-[ethyl [(heptadecafluorooctyl)-sulfonyl]amino]-ethylmethacrylate) (FAB) is a fluorinated acrylic polymer that was synthesized in one step from the monomer (Aldrich). The monomer (100 g, mmol) was dissolved into α,α,α-trifluorotoluene (300 g, mmol) and activated charcoal was added. The activated charcoal was removed by vacuum filtration. The solution was added to a four-neck round bottom flask with 0.8% by monomer weight lauroyl peroxide. The flask was equipped with a mechanical stirrer, a sparger, and reflux condenser, and a refrigeration unit to cool the condenser. The flask was purged with prepure nitrogen at a flow rate of 61 mL min^{-1}. The polymerization was carried out at 72 °C for 48 hours, and then was heated to 100 °C for 8 hours. It was observed that the solution became very viscous during the polymerization.

The polymer was purified and isolated by precipitation. The polymer solution was diluted with 200 mL of α,α,α-trifluorotoluene. The resulting solution was then added dropwise to 1,200 mL solution of 60% hexane, 40% methylene chloride at 0 °C under constant agitation. The resulting powder was collected by vacuum filtration and dried under vacuum.

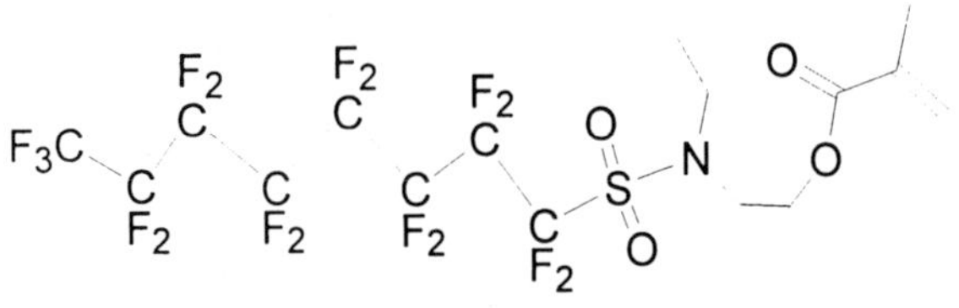

Figure 2. Chemical structure of 2-[ethyl[(heptadecafluorooctyl)-
sulfonyl]amino]-ethylmethacrylate

General procedure for synthesis of osmium complexes.[5, 6]

The osmium complexes were synthesized by reacting 1.000g (2.08 mmol) of $(NH_4)_2OsCl_6$ with 2.05 equivalents of polypyridyl (N-N) ligand in 25 mL of refluxing DMF (Aldrich) under inert atmosphere for 3 hours. The resulting solution was filtered, washed with DMF, cooled to 0 °C, and then added dropwise to a water solution of sodium dithionite (2.00g in 400 mL) at 0 °C. The resulting purple precipitate of $Os(N-N)_2Cl_2$ was filtered and washed with deionized water. $Os(N-N)_2Cl_2$ was reacted with 1.05 eq. of cis-1,2vinylenebis(diphenylarsine) (dpaene), cis-1,2-vinylenebis(diphenylphosphine) (dppene), methylenebis(diphenylarsine) ligand in a refluxing mixture of 2,2'-ethoxyethoxyethaonol (Aldrich) and glycerol (75:25 by volume) for 2 hours under inert atmosphere. The complexes were precipitated by dropwise addition to a water solution of KPF_6 and filtered. Complexes were then purified on silica using either toluene/acetonitrile or acetonitrile/water/KPF_6 solvent systems. The structures of the resulting complexes are shown in Figure 3. Elemental analysis:

Complex 1: Calculated: C, 54.21; H, 3.37; N, 3.46. Found: C, 54.82; H, 3.09; N, 3.65.

Complex 2: Calculated: C, 54.48; H, 3.46; N, 3.43. Found: C, 54.77; H, 3.55; N, 3.19.

Complex 3: Calculated: C, 51.63; H, 4.03; N, 4.15. Found: C, 51.39; H, 4.00; N, 4.11.

Complex 4: Calculated: C, 48.47; H, 3.79; N, 3.90. Found: C, 48.53; H, 3.75; N, 3.95.

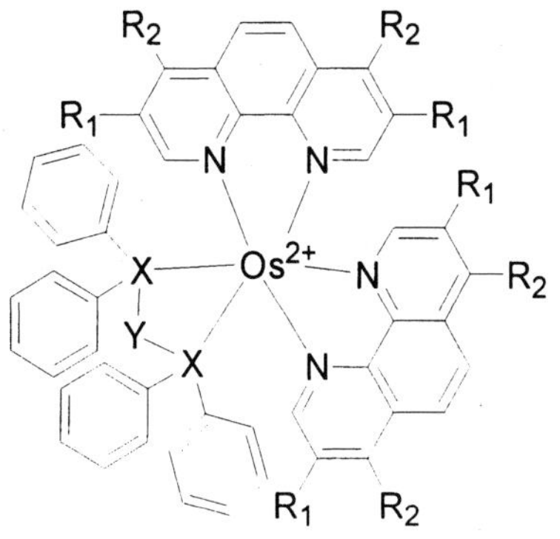

Complex	X	Y	R1	R2
1	As	CH_2	H	C_6H_5
2	As	$CH_2 CH_2$	H	C_6H_5
3	P	C=C	CH_3	CH_3
4	As	C=C	CH_3	CH_3

Figure 3. Chemical structures of the various osmium complexes being reported.

Application and Characterization of PSP

Osmium complex (1.0 mg) was added to a 25 mL vial and 2.0 ml of acetone was added to dissolve the complex. To this was added 1.0g of FAB and 20.0 mL of α,α,α-trifluorotoluene. Both the polymer and the complexes were readily soluble in the solvent.

One-inch square aluminum plates were scrubbed and polished with abrasive pads and then rinsed with acetone and methylene chloride. After allowing it to completely dry, it was then spray-painted with the pressure paint, using nitrogen as the propellant. After air-drying, the samples were placed into vacuum oven and heated to 100 °C for thirty minutes.

Stern-Volmer plots, temperature dependence, and photo-degradation were obtained using the "PMT survey apparatus." This apparatus simultaneously monitors, and controls pressure, temperature, and luminescence intensity. A regulated tungsten-halogen lamp provides the illumination source. Pressure may be varied from millitorr to 1 atm pressure and the temperature may be varied from 5 °C to 50 °C. Photo-stability studies were conducted at constant pressure and temperature. A full description and functionality of the survey apparatus has been published in our previous reports.[7]

Results and Discussion

The osmium complexes 1 through 4 have been incorporated into PSP with FAB and various physical properties have been measured. These results are compared to [ruthenium tris(4,7-diphenyl-1,10-phenanthroline)]$^{2+}$ 2PF$_6^-$ (rubth) (GFS Chemicals) and was recrystallized from toluene/acetonitrile prior to use. Rubth is a standard dye often used for PSP. The rubth/FAB based PSP gave a Stern-Volmer dynamic range of 55% (Figure 4) as defined as 1 − (emission intensity at 1 atm pressure / emission intensity at vacuum). Thus the emission intensity from the rubth dye changed by over a factor of 2 when the pressure is reduced from atmospheric to vacuum. Under illumination at 400 nm (fwhm = 20 nm) with a power density of 700 μw cm^{-1}, the rubth dye exhibited photo-degradation at the rate of 6% per 90 minutes (Figure 5.). The temperature dependence was observed to be 1.4% ^{o}C^{-1} (Figure 5).

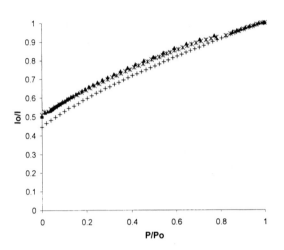

Figure 4. Stern-Volmer plot measured at 25 ^{o}C for rubth(+), complex 3(▲) and complex 4(∗).

The osmium complexes typically exhibited much less photo-degradation and temperature dependence than the rubth dye. The properties observed may be highly dependent upon the ligands chosen to form the complexes. Complexes 1 and 2 were based upon the 4,7-diphenyl-1,10-phenanthroline ligand structure found in the rubth dye. Complex 1 was synthesized with the bis(diphenylarseno) methane ligand and complex 2 with the 1,2-bis(diphenylarseno) ethane ligand. These complexes featured a variety of absorption bands between 250 nm and 550 nm. The bands that appear at ~280 nm are due to the ligand centered π–π*transitions and have an ε of 65,000 L cm^{-1} mol^{-1}. At lower energy appear the spin allowed charge transfer bands (390 nm, ε 19,000 L cm^{-1} mol^{-1}), and the spin forbidden charge transfer bands (525 nm, ε 4,000 L cm^{-1} mol^{-1}). The emission peak of complex 1 occurred at 645 nm with a quantum yield of 6% and complex 2 at 648 nm with a quantum yield of 12%. The emission lifetime of these complexes was measured to be 1,500 ns. The pressure dependence of the complexes was measured to give a dynamic range of 30%, but the two were very

113

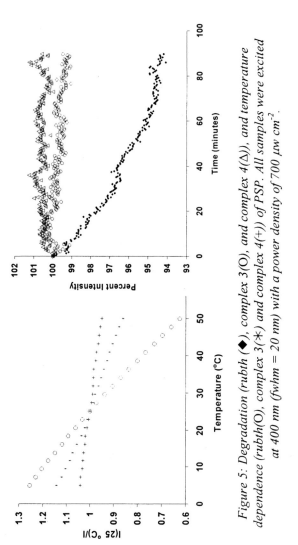

Figure 5: Degradation (rubth (◆), complex 3(O), and complex 4(△)), and temperature dependence (rubth(O), complex 3(✳) and complex 4(+)) of PSP. All samples were excited at 400 nm (fwhm = 20 nm) with a power density of 700 μw cm⁻².

different in their photo-degradation and temperature dependence. The emission of complex 1 was observed to decrease 0.33 % $^{\circ}C^{-1}$ rise in temperature, and degraded nearly 14% after 90 minutes of illumination at 400 nm. Under the same circumstances, complex 2 was observed to have temperature dependence of 0.1% $^{\circ}C^{-1}$ (Figure 6), and degraded 1.5% (Figure 7). The temperature dependence measured was much lower than that of the rubth dye, while the degradation of complex 1 was significantly higher and complex 2 was significantly lower. The dynamic range of complexes 1 and 2 were much narrower than the rubth dye. This may be due to the short-lived excited states for these osmium complexes. The lifetime of the rubth dye has been reported at 6 μs,[4] while the osmium complexes were much faster.

Complexes 3 and 4 were based upon the 3,4,7,8-tetramethyl-1,10-phenanthroline ligand. Complex 3 was synthesized with cis-1,2-bis(diphenylphosphino) ethene and complex 4 with cis-1,2-vinyl-enebis(diphenylarsine) ligand. The ligand-centered transition appeared at 272 nm, with the charge transfer absorption bands being observed at 367-377 nm (ε of 13,000 - 17,000 L cm^{-1} mol^{-1}) for the spin allowed bands and 441-450 nm (ε of 4,000 – 5,000 L cm^{-1} mol^{-1}) for the spin forbidden bands. Emission for complex 3 was observed at 572 nm and the lifetime was measured to be 3,400 ns. Emission for complex 4 was observed at 588 nm and the lifetime was measured to be 2,800 ns. The faster emission rate observed for complex 4 vs. complex 3 may have been due to the presence of the arsenic atoms. Arsenic, being heavier than phosphorus, has a larger spin orbit-coupling constant, which relaxes the rules governing changes in spin. This usually leads to faster rates of intersystem crossing to the triplet state leading to faster phosphorescence. Both complexes 3 and 4 were observed to give a dynamic range of 50% for PSP based upon them (Figure 4). Both exhibited very low rates of photo-degradation, with complex 4 giving no photo-degradation, and complex 3 0.8% per 90 minutes of illumination (Figure 5). The emission form complex 4 was observed to decrease 0.2% per degree centigrade rise in temperature (Figure 5). Under the same conditions, complex 3 was observed to decrease 0.84% $^{\circ}C^{-1}$ (Figure 5). Both of these were much lower than the 1.4% $^{\circ}C^{-1}$ decrease observed for the rubth dye while giving a similar dynamic range (Figure 5).

Of the observed properties, complex 1 was observed to exhibit higher temperature dependence and had the fastest rate of photo-degradation, even faster than the rubth dye. This result was unexpected, as it was expected that the osmium complexes would be less temperature dependent than the corresponding ruthenium complexes. As osmium is in the third row, the crystal field splitting for complexes with the metal are larger than for complexes formed with ruthenium, which is in the second row. The larger crystal field splitting has a tendency to lead to higher energy metal centered states for osmium, thus reducing temperature dependence and photo-degradation. Usually the metal

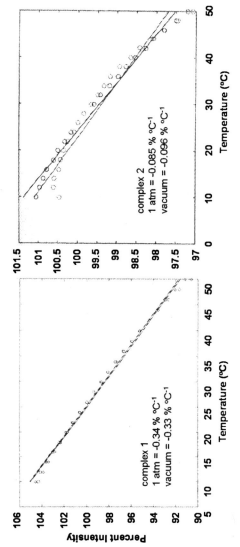

Figure 6. Temperature dependence of PSP based upon complexes 1 and 2.

116

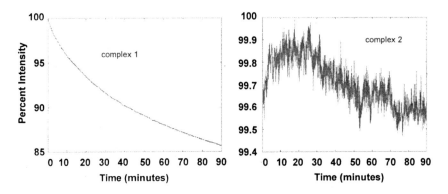

Figure 7. Photo-degradation of complexes 1 and 2 after 90 minutes of illumination at 400 (fwhm = 20 nm) nm with a power density of 700 μw cm^{-2}.

centered states are Laporte forbidden due to the like symmetry with the ground state. However, under certain circumstances the metal centered state may be lowered in energy and become more easily populated. For complexes in octahedral symmetry, small perturbations may lead to relaxation of the Laporte rules. Other molecules, such as solvent molecules, may cause the perturbations if the complex is dissolved, colliding with the complex, slightly distorting it.[8] In the case of complex 1, the distortion to the **d** orbitals may be inherent with the molecule. The methane bridge may cause a significant bending of the **d** orbitals of the osmium due to a narrow bite angle. This would have an effect of lowering the energy of the MC state, and there would be significant ring strain of the four-member $Os-As_2-C$ ring. The lowering of the MC state would make the state more accessible with increasing temperature. Together the observation of greater temperature dependence, and the presence of ring strain would increase the likelihood of ligand loss. Thus, the observed increases in photo-degradation and temperature dependence were observed.

Complexes 3 and 4 were observed to give broader dynamic range than complexes 1 and 2. This may be due to the longer-lived excited states for complexes 3 and 4, which were almost a factor of 2 longer than complexes 1 and 2. The pressure paint functions by a bi-molecular collision between oxygen and the triplet excited state of the complex. When this takes place, an energy transfer occurs between the excited complex and the triplet ground state of oxygen. The result of the reaction is the production of singlet oxygen and the return of the complex to the singlet ground state. There is a greater chance of collision of oxygen with the excited state of complexes 3 and 4 since they have a longer lifetime; thus, increasing dynamic range of the PSP based on these complexes.

Conclusion

Several osmium complexes have been synthesized and incorporated into PSP. The PSP's featured dynamic ranges up to 50%, which were similar to ruthenium tris(bathophenanthroline). Temperature dependence was as low as 0.1% $°C^{-1}$ and very low to non-existent rates of photo-degradation was achieved. Complexes based upon 3,4,7,8-tetramethyl-1,10-phenanthroline exhibited the broadest dynamic ranges due to the longer lived excited states of the complexes formed with this ligand.

Acknowledgments

The authors wish to thank the AFOSR Smart Skin MURI program for funding this research.

References

1. (a) Puklin, E.; Carlson, B.; Gouin, S.; Costin, C.; Green, E.; Ponomarev, S.; Tanji, H.; Gouterman, M., J. *App. Poly. Sci.* **2000**, *77(13)*, 2795. (b) Ji, H-F.; Shen, Y.; Hubner, J. P.; Carroll, B. F.; Schmehl, R. H.; Simon, J. A.; Schanze, K. S., *Appl. Spect.* **2000**, *54(6)*, 856.(c) Gouin, S.; Gouterman, M., J. *App. Poly. Sci.* **2000**, *77(13)*, 2815. (d) Khalil, G. E.; Costin, C.; Crafton, J.; Jones, G.; Grenoble, S.; Gouterman, M.; Callis, J. B.; Dalton, L. R., *Sens. and Actu., B: Chem.* **2004**, *B97(1)*, 13.
2. (a)Vander Donckt, E.; Camerman, B.; Hendrick, F.; Herne, R.; Vandeloise, R., *Bull. Soc. Chim. Belg.* **1994**, *103*, 207. (b) Amao, Y.; Ishikawa, Y.; Okura, I., *Anal. Chim. Acta* **2001**, *445*, 177. (c) Gao, R.; Ho, D. G.; Hernandez, B.; Selke, M.; Murphy, D.; Djurovich, P. I.; Thompson, M. E., *J. Am. Chem. Soc.* **2002**, *124*, 14828.(d) Carlson, B.; Khalil, G.; Gouterman, M.; Dalton, L., *Polym. Prepr.* **2002**, *43*, 590.(e) DeRosa, Maria C.; Hodgson, Derek J.; Enright, Gary D.; Dawson, Brian; Evans, Christopher E. B.; Crutchley, Robert J., *J. Amer. Chem. Soc.* **2004**, *126(24)*, 7619.
3. (a) Jiang, F.; Xu, R.; Wang, D.; Dong, X.; Li, G., *Gongneng Cailiao* **2000**, *31(1)*, 72.(b) Schanze, K. S.; Carroll, B. F.; Korotkevitch, S., Morris, M., *AIAA J.* **1997**, *35(2)*, 306.(c) Jiang, F-Z.; Xu, R.; Wang, D-Y.; Dong, X-D.; Li, G-C.; Zhu, D-B., Jiang, L., *J. Mat. Res.* **2002**, *17(6)*, 1312. (d) Sakaue, H.; Gregory, J. W., Sullivian, J. P., *AIAA J.* **2002**, *40(6)*, 1094. (e) Bowman, R. D.; Kneas, K. A.; Demas, J. N.; Periasamy, A., *J. of Microscopy* **2003**,

211(2), 112. (f) Schanze K. S.; Carroll B. F.; Korotkevitch S.; Morris M., *AIAA J.* **1997**, *35(2)*, 306. (g) Mingoarranz, F. J.; Moreno-Bondi, M. C.; Garcia-Fresnadillo, D.; de Dios, C.; Orellana, G., *Mikrochimica Acta* **1995**, *121(1-4)*, 107-118.

4. Juris, A.; Balzani, V.; Barigelletti, F.; Campagna, S.; Belser, P.; Zelewsky, A., *Coordina. Chem. Rev.* **1988**, *84*, 85.

5. Kober, E.; Caspar, J.; Sullivan, P.; Meyer, T., *Inorg. Chem.* **1988**, *27*, 4587.

6. Kober, E. M.; Marshall, J. L.; Dressick, W. J.; Sullivan, B. P.; Caspar, J. V.; Meyer, T. J., *Inorg. Chem.* **1985**, *24*, 2755.

7. Baron, A. E. University of Washington Doctorial Thesis: On time- and spatially-resolved measurements of luminescence-based oxygen sensors (pressure sensitive paint). **1996**, 201 pp.

8. Huheey, J. E., Keiter, E. A., Keiter R. L., "Inorganic Chemistry: Principles of Structure and Reactivity", 4th Edition, Harper Collins, New York, **1993**, 438.

Novel Coatings

Chapter 9

Development of a Removable Conformal Coating through the Synthetic Incorporation of Diels–Alder Thermally Reversible Adducts into an Epoxy Resin

J. H. Aubert[1], D. R. Tallant[2], P. S. Sawyer[1], and M. J. Garcia[2]

[1]Organic Materials and [2]Materials Characterization Departments, Sandia National Laboratories, Albuquerque, NM 87185

An epoxy-based conformal coating with a very low modulus has been developed for the environmental protection of electronic devices and for stress relief of those devices. The coating was designed to be removable by incorporating thermally-reversible Diels-Alder (D-A) adducts into the epoxy resin utilized in the formulation. The removability of the coating allows us to recover expensive components during development, to rebuild during production, to upgrade the components during their lifetime, to perform surveillance after deployment, and it aids in dismantlement of the components after their lifetime. The removability is the unique feature of this coating and was characterized by modulus versus temperature measurements, dissolution experiments, viscosity quench experiments, and FTIR. Both the viscosity quench experiments and the FTIR measurements allowed us to estimate the equilibrium constant of the D-A adducts in a temperature range from room temperature to 90 °C.

Introduction

The use of conformal coatings in electronic applications, (Printed wiring boards; PWBs), can provide component protection from a variety of hostile environments. Conformal coatings provide a secure envelope around a circuit board and its components and act as a barrier against chemicals, moisture, fungus, dust and other environmental contaminants. There is an additional need at Sandia National Laboratories for a conformal coating that is relatively easy to remove when required. Removal of the conformal coating from PWBs allows us to repair or upgrade the electronic assembly without causing damage to the board or its components. Additionally, non-damaging removal of the conformal coating is advantageous for surveillance and for dismantlement. There are few commercial conformal coatings that are removable.

An epoxy-based conformal coating with a very low modulus has been developed for the environmental protection of electronic devices and for stress relief of those devices. The coating was designed to be removable by incorporating thermally reversible Diels-Alder (D-A) adducts into the epoxy resin utilized in the formulation of the coating. Figure 1 shows the reported reversibility of the D-A adduct between a maleimide and a furan and the approximate temperatures where the reversion is reported to occur (1). The removability is the unique feature of this coating and was characterized by resin viscosity, modulus versus temperature measurements and dissolution experiments on conformal coating, and Infrared (IR) measurements on resin. IR was used to monitor the thermally reversible nature of the D-A adducts. IR measurements coupled with chemometrics allowed us to estimate the equilibrium constant of the D-A adducts in a temperature range from room temperature to 90 °C.

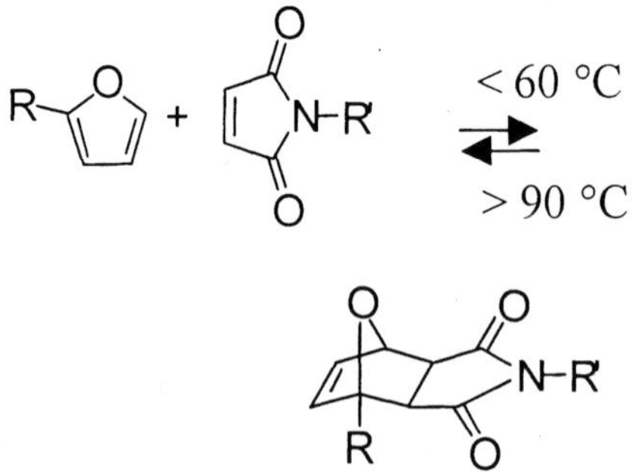

Figure 1. Thermally reversible Diels-Alder adduct formation between a Furan and a Maleimide according to reference 1.

Experimental

Materials

Technical grade nonylphenol, furfuryl glycidyl ether (96%) and furfuryl alcohol (99%) were obtained from Aldrich Chemical Co., (Milwaukee, WI). Diamine curatives, polyoxypropylenediamines (Jeffamine® D-230 (DP~2.86) and Jeffamine® D-2000 (DP~34.2)) were obtained from Huntsman Chemical Co., (Houston, TX). Epon™ 8121 was obtained from Resolution Performance Products, (Houston, TX). All reagents were used as received.

Viscosity and Dynamic mechanical analysis (DMA)

Viscosity and DMA data was collected using a Rheometrics Scientific SR5 Stress Rheometer (now TA Instruments, New Castle, DE). Dynamic testing was performed on cured disks of conformal coating having a diameter of 25 mm and a thickness of 2 mm, at a frequency of 1 Hz and with the temperature ramped at a rate of 3 °C/min. Prior to the test, the cured sample was heated to softening and the normal force adjusted until full contact between the sample and plates was obtained. Resin viscosity was measured at various temperatures with parallel plate geometry (25 mm diameter, maximum 0.5 mm thickness) and in dynamic mode. The bottom plate was a Peltier that allowed rapid temperature jumps.

FTIR/chemometrics

Infrared (IR) spectra were obtained in transmission using a Fourier Transform Infrared spectrometer. The resins were pressed between salt windows and heated in a controlled temperature cell. Multivariate chemometric techniques were used to analyze the IR spectra. These techniques decompose a set of spectra into a set of factors, which can be thought of as the components common to the set of spectra. The relative contribution of each factor to a spectrum is the score for that spectrum. The controlling algorithm is that the sum of the set of factors, scaled by each factor's associated score for an experimental spectrum, should approximate that spectrum. Analysis of the factors identifies components or functional groups associated with them. The scores for a factor reflect how the relative amounts of the components or functional groups associated with that factor change from spectrum to spectrum. Thus, comparison of factors (components) with their scores (relative contributions to the spectra) shows how the components vary in concentration in the materials from which the set of spectra were obtained. From this analysis, quantitative measures of the degree of completion of the reaction shown in Figure 1 were obtained.

Resin Synthesis

Octamethyltetrasiloxane-di-4,1-phenoxy bismaleimide was prepared according to a previously reported procedure (2). Proton NMR spectra indicated that this bismaleimide was prepared with approximately 90% purity. The diepoxy resin and the dialcohol used in this study were prepared by neat reaction of Octamethyltetrasiloxane-di-4,1-phenoxy bismaleimide with furfuryl glycidyl ether or furfuryl alcohol respectively at 60 °C as shown in Scheme 1.

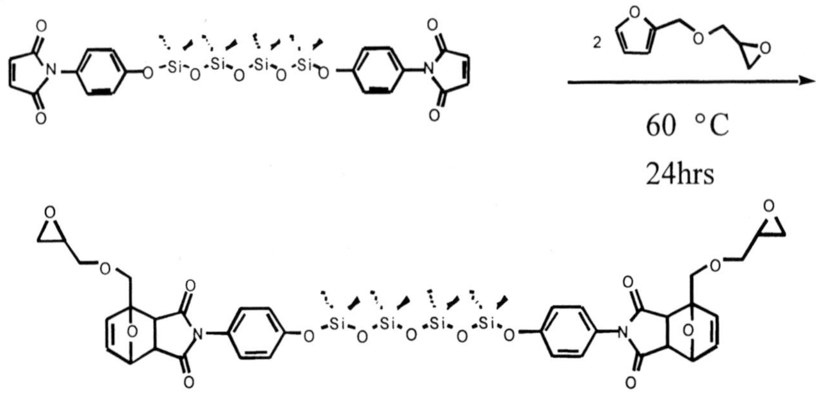

Scheme 1. Preparation of diepoxy resin, RER 1. The dialcohol is prepared in an analogous manner by substituting furfuryl alcohol for furfuryl glycidyl ether.

Results and Discussion

Conformal Coating Formulation

The formulation of the removable conformal coating, RCC200, is shown in Table I. The primary drivers for this particular formulation were a requirement of a low modulus and removability. Many properties (thermal, electrical, mechanical) of this formulation have been determined but will not be reported here. Thermal cycling experiments with this coating on populated boards has been studied and will be reported elsewhere. The application of the coating was done by first degassing the material and then dispensing with a syringe. The epoxy cure was done at room temperature for 16 hours followed by 2 hours at 60 °C.

DMA

Figure 2 shows a plot of the shear storage modulus, G', obtained from DMA testing as a function of temperature for three temperature ramps performed on the same disk of RCC200 (cured conformal coating). During the first temperature ramp, the modulus fell beginning at about 100 °C by more than 95% as the retro D-A reaction was expected to occur as shown in Figure 1. When the sample was cooled, the modulus recovered as the D-A adducts were expected to reform. This process was reversible for numerous cycles. During the second temperature ramp, the modulus was very close to that measured during the first ramp. Again, above about 100 °C, the modulus fell dramatically. After the second temperature ramp, the sample was cooled and the modulus recovered. Some contact between the sample and the plates was then lost and the measurement of modulus became somewhat erratic. However, this experiment demonstrates the thermally reversible nature of the D-A adducts as manifested by the thermally reversible shear storage modulus of the conformal coating.

The modulus at temperatures below 100 °C is typical for an elastomeric adhesive, such as Sylgard® 184. The modulus of this cross-linked elastomer does not vary significantly with temperature (3).

Table I. Formulation of RCC200, a removable conformal coating.

PART A Epoxy resins	Percent of PART A	PART B Curatives	Percent of PART B	Weight of PART B per 100 gm of PART A
RER1	75%			
Epon™ 8121	25%			
Total PART A	100%			
		Jeffamine® D-230	12.6%	10.4
		Jeffamine® D-2000	71.8%	59.3
		Nonyl phenol	15.6%	12.9
		Total PART B	100%	82.6

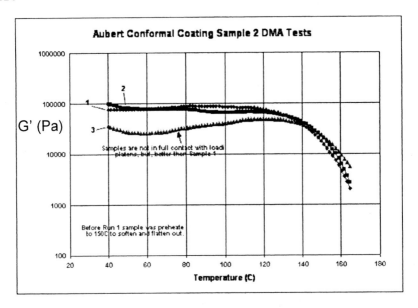

Figure 2. Shear modulus, G' (Pa), of RCC200 as a function of temperature (°C). The reversible softening of the material is shown by repeated heatings (1-3). (1 psi = 6895 Pa). Some contact between the sample and plates was lost on run 3 resulting in the observed lower modulus.

FTIR/chemometrics

RCC200 (Table 1) was too complicated to observe the reversible D-A adducts by IR. For simplicity, we initially performed IR on the diepoxy resin, RER 1, shown in Scheme 1. However, at elevated temperatures, the diepoxy would self-polymerize and this also complicated interpretation of the spectra. To avoid this, we prepared the dialcohol formed from the bismaleimide and furfuryl alcohol (also described in Scheme 1). We assumed that the reversible nature of the D-A adduct in the dialcohol is very similar to that in the diepoxy. For interpretation of the behavior of the cured conformal coating, we also assumed that the reversible nature of the D-A adduct was similar in the cured resin as it was in the uncured resin. The IR spectra of the initial, nonreacted, mixture of furfuryl alcohol and octamethyltetrasiloxane-di-4,1-phenoxy bismaleimide at ambient is shown in Figure 3. Also shown is the IR spectra of this mixture after adduct formation at 60 °C for 2991 min. Most of the change in the IR spectra occurs in the first two hours. Spectra at intermediate times were also obtained but are not shown here. IR spectra of the individual starting components was obtained but is also not shown.

FTIR/chemometrics showed that a reversible reaction did occur in the dialcohol upon heating and cooling between room temperature and 90 °C for numerous cycles. Figure 4 shows the spectral region that had the largest thermally reversible changes. The absorbance at 1191 cm^{-1} corresponds to the C-O-C/C-N-C bonds of adducts, whereas the absorbance at 1150 cm^{-1} is due to the corresponding C-O-C/C-N-C bonds of the maleimide and furan.

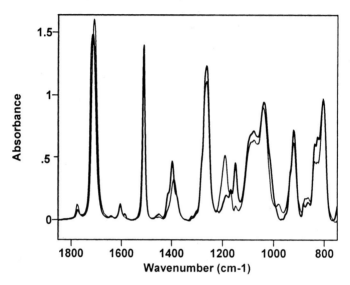

Figure 3. The solid plot shows the absorbance of the initial mixture of bismaleimide and furfuryl alcohol at room temperature prior to adduct formation. The dashed plot shows the absorbance of the mixture after adduct formation at 60 °C for 2991 min.

By assuming that the equilibrium of the reversible reaction was 100% adduct (Figure 1) at room temperature, an extent of reaction could be determined at each of the elevated temperatures where measurements were performed. The extent of reaction for the reversible D-A adduct formation (Figure 1) can then be related to an equilibrium constant as shown in equation (1).

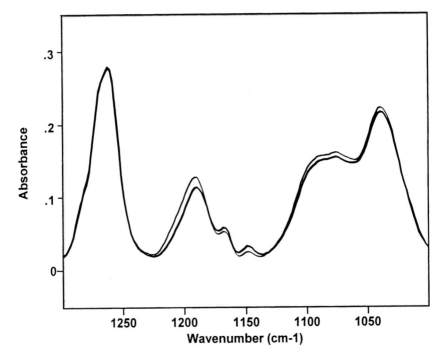

Figure 4. Absorbance of the adduct (fully formed initially at 60 °C) at ambient temperature (dashed line) and at 90 °C (solid line). Absorbance at 1150 cm^{-1} corresponds to C-O-C/C-N-C bonds in precursors and at 1191 cm^{-1} corresponds to C-O-C/C-N-C bonds in adduct. The temperature dependence of the spectra was reversible.

A is the activity of the species in brackets. A best fit to the data gave: K(T) = 7.2×10^{-11} exp{9787/T}, where T is in °K. Predictions from this equation for the temperature range studied and also extrapolated to some higher temperatures are shown in Table II.

Viscosity

Figure 5 shows how the viscosity increases as D-A adducts form from the starting materials (bismaleimide and furfuryl alcohol) at 60 °C. Most of the viscosity rise occurs within the first two hours with a slower increase in viscosity occurring over 48 hours. This is consistent with the IR measurements.

Table II. Estimated equilibrium constants based upon IR measurements on a diol containing two D-A adducts.

Temperature	Equilibrium Constant
23 °C	16187
60 °C	413
90 °C	36
120 °C	5
150 °C	1

The viscosity of some equilibrated stoichiometric mixtures of furfuryl alcohol and bismaleimide have been measured at various temperatures. The viscosity decreases significantly with increasing temperature due in part to the retro D-A reaction.

With the capabilities of the Peltier plate on the rheometer, we were able to do quench experiments on solutions and obtain an estimate of adduct equilibrium. The Peltier plate could be quenched from 90 °C to 60 °C in approximately 30 seconds. The temperature of a solution between the plates, at the gaps utilized, could be equilibrated within approximately 1.5 min. This is much faster than that the adduct formation (hours) as shown in Figure 5. Therefore, we were able to quench a solution equilibrated at 90 °C to 60 °C and measure the initial viscosity (related to the extent of adduct formation at 90 °C) and follow the viscosity rise as adducts formed and approached the equilibrium level at 60 °C. Such a quench experiment is shown in Figure 6.

After temperature equilibration (approximately 1.5 min), the initial viscosity was approximately 11 Pa-s. This compares to the equilibrium viscosity at 60 °C of approximately 31 Pa-s. The difference corresponds to the difference in adduct equilibrium at 90 °C and at 60 °C. If the relationship between adduct equilibrium and viscosity were established, then the difference in adduct equilibrium at 90 °C and 60 °C could be determined. This adduct equilibrium/viscosity relationship is expected to be nonlinear.

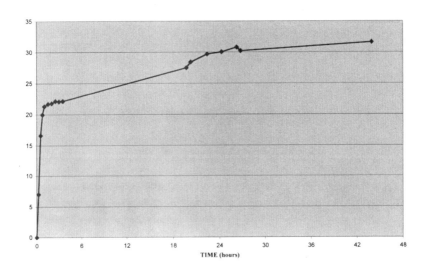

Figure 5. Viscosity (Pa-sec) increase of the stoichiometric mixture of furfuryl alcohol and the bismaleimide at 60 °C as a function of time (min) for adduct formation.

Although this relationship is not known, we estimated two points. We measured viscosities of non-stoichiometric mixtures of furfuryl alcohol and bismaleimide by mixing either 15% or 20% excess bismaleimide. When equilibrated these mixtures have a viscosity at 60 °C of 27 Pa-s (15% excess) or 12 Pa-s (20% excess). These solutions differ from quenched solutions because they are missing the excess furfuryl alcohol that corresponds to the excess bismaleimide. If the excess furfuryl alcohol were present, the visocosity would be somewhat lower.

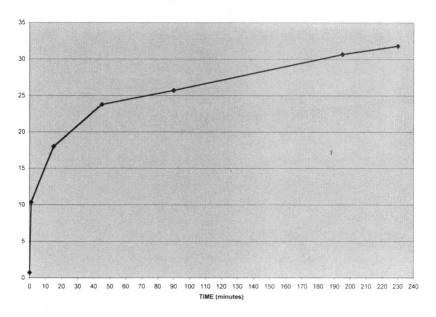

Figure 6. Viscosity (Pa-sec) rise in a stoichiometic mixture of furfuryl alcohol and bismaleimide equilibrated at 90 °C and then quenched to 60 °C as a function of time (min) for adducts to reform.

The 20% excess value is close to the value of viscosity measured after the quench, 11 Pa-s, and implies that at 90 °C approximately 20% more of the adducts are nonreacted (open) than at 60 °C. This is consistent with the FTIR results described above. FTIR indicated that at 60 °C approximately 5% of adducts were open (K = 412), and at 90 °C approximately 15% of adducts were open (K = 36) – a 10% difference. The viscosity and FTIR are therefore consistent. The viscosity numbers are higher partly due to the higher viscosity measured without the presence of excess furfuryl alcohol in the non-stoichiometric mixtures.

Removability Experiments

We have removed cured conformal coating with two approaches. One approach is to raise the temperature to 90 °C, where a reasonable fraction of adducts are open, and to expose the coating to a mild solvent (e.g. 1-butanol). Such conditions do not damage our electronics but allow the coating to dissolve. The data in Table 2 indicate that at 90 °C, the adduct equilibrium constant is 36, which corresponds to approximately 15% of adducts open. Our proposed mechanism is the following. If solvent is present, an opened D-A adduct will be solvated and remain open. Over time, more and more of the adducts open, are solvated, and the coating slowly dissolves. A 10 mil thick coating can be dissolved in approximately 2 days in this manner. This is an acceptable time for our intended purposes.

The second approach is to utilize a removal solvent that partakes in the adduct equilibrium. For this we use furfuryl alcohol. Removal experiments were performed at 50 °C. With a large excess of furfuryl alcohol, the D-A equilibrium in the resin material, can be pushed in the direction of very little adduct formed with resin. Although the D-A adducts will still exist, they will be dominated by those formed with furfuryl alcohol, rather than the furan within the resin, because of the high concentration of furfuryl alcohol compared to resin. Our electronics survived furfuryl alcohol at 50 °C and a 10 mil thick coating can be dissolved in approximately 2 days.

Conclusions

A removable conformal coating was developed to satisfy a number of criteria including having a low modulus and having an ability to be removed without causing damage to coated electronics. Coating removal was demonstrated by dissolution experiments performed in 1-butanol at 90 °C and in furfuryl alcohol at 50 °C. The removal (dissolution) was possible because of the incorporation of thermally reversible D-A adducts into the epoxy resin utilized in the coating formulation. The retro D-A reaction was demonstrated to be reversible on a resin similar to the one used in the coating by the use of IR/chemometrics and by viscosity quench experiments. The retro D-A reaction and the coating removability were shown to be consistent with the temperature dependence of the shear storage modulus. Significant modulus loss was observed at 150 °C where the fit to the IR/chemometrics indicated an adduct equilibrium constant of approximately one.

Acknowledgements

Sandia is a multi-program laboratory operated by Sandia Corporation, a Lockheed Martin Company, for the United States Department of Energy's

National Nuclear Security Administration under contract DE-AC04-94AL85000.

References

1. Stevens, M. P., *Polymer Chemistry An Introduction*, 3rd edition, Oxford University Press, NY, **1999**.
2. McElhanon, J. R., Russick, E. M., Wheeler, D. R., Loy, D. A., and Aubert, J. H., *J. Appl. Polym. Sci.*, **2002**, *85*, 1496-1502.
3. Aubert, J. H., *J. of Adhesion*, **2003**, *79*, 609-616.

Chapter 10

Electroactive Polymer Coatings as Replacements for Chromate Conversion Coatings

P. Zarras[1], J. He[2], D. E. Tallman[2], N. Anderson[1], A. Guenthner[1], C. Webber[1], J. D. Stenger-Smith[1], J. M. Pentony[3], S. Hawkins[1], and L. Baldwin[1]

[1]NAWCWD, Polymer Science and Engineering Branch (Code 498200D), 1900 North Knox Road (Stop 6303), China Lake, CA 93555–6106
[2]Department of Chemistry, North Dakota State University, Fargo, ND 58105–5516
[3]NAWCWD, Physics and Computational Sciences Branch (Code 498100D), 1900 North Knox Road (Stop 6303), China Lake, CA 93555–6106

NAVAIR-WD has successfully synthesized an electroactive polymer, poly(2,5-bis(N-methyl-N-hexylamino)phenylene vinylene), (BAM-PPV) in high yield and purity. BAM-PPV has also been characterized using advanced spectroscopic, electrochemical and thermal analysis techniques (NMR, FTIR, ENM, TMA, DSC and TGA) to determine its structure, corrosion prevention mechanism, mechanical and thermal properties, respectively. BAM-PPV coated aluminum samples have repeatedly survived 336 hours neutral salt fog spray exposure (ASTM B117), which is a current military requirement for alternatives to chromate conversion coating (CCC) pretreatments. Testing the polymer system as a conversion coating under neutral salt fog conditions with full military coatings has also been performed. The scribed BAM-PPV Al 2024-T3 coupons with a non-chromate primer and topcoat showed evidence of corrosion at 2000 hours. Coupons with BAM-PPV as the pretreatment with a chromated primer and a topcoat showed no corrosion along the scribed areas after 2000 hours of salt fog exposure. These were compared to a fully chromated system which also showed, no corrosion along the scribed area at 2000 hours of exposure.

Introduction

Chromate conversion coatings (CCC) are applied via immersion or spraying onto both aluminum and steel substrates. These coatings provide both corrosion inhibition and adhesion promotion between the primer and the substrate. (*1*). Several recent studies have shown that residual hexavalent chromium (Cr(VI)) in chromate conversion coatings provides corrosion protection via a self-healing mechanism (*2,3*). However, Cr(VI) is a known carcinogen (*4*) and is highly regulated by the Environmental Protection Agency (EPA) and Occupational Safety and Health Agency (OSHA) (*5*).

Any viable alternative to Cr(VI) coatings must meet or exceed the performance of Cr(VI) (*1-3*). Ideally, these alternative coatings must be able to passivate the metal surface allowing the corrosion current to shift to the noble metal region (*6*). Several studies over the past decade have focused on electroactive polymers (EAPs) as corrosion preventive coating (*7-10*). These studies have focused almost exclusively on polyaniline (PANI) as a primer and have confirmed a passivation mechanism for corrosion inhibition (*11-13*). More recent studies have focused on PANI as a replacement for chromated primers (*14,15*). Several studies using polythiophenes and polypyrroles have been used to coat both ferrous and nonferrous substrates (*16,17*). Electrochemical studies were performed on these EAPs as primer replacements (*18*), but very little work has been done on alternative pretreatments using EAPs. Scientists at the NAVAIR-WD have successfully synthesized a new poly(*p*-phenylene vinylene) (PPV) derivative that may be an alternative to chromated pretreatments. This compound poly(2,5-bis(N-methyl-N-hexylamino)phenylene vinylene, BAM-PPV has shown corrosion prevention in simulated seawater (*19-21*). BAM-PPV in its undoped form has a measured conductivity of 10^{-11} S/cm. When pressed pellets of BAM-PPV were doped with iodine a conductivity of 10^{-6} S/cm was measured. BAM-PPV was studied using various techniques to understand its thermal stability and interaction with aluminum substrates. Its corrosion prevention properties were also evaluated using accelerated weathering conditions to determine its effectiveness as an alternative pretreatment to CCC.

Experimental

The monomer, (2,5-bis(chloromethyl)-4-(hexamethylamino)-phenyl)-hexamethylamine dihydrochloride) was prepared according to published work (*22*). BAM-PPV solutions were prepared and thin films of BAM-PPV were coated onto aluminum substrates (Al 2024-T3) using an airbrush technique (film thickness approximately ~1.5 μm). Differential scanning calorimetry (DSC) was performed using a TA Instruments 2910 Differential Scanning Calorimeter at a heating rate of 10°C/min under nitrogen. Thermogravimetric analysis (TGA) was performed using a TA Instruments 2950 TGA scanning at 5°C/min (under nitrogen and air). ^{1}H and ^{13}C NMR data was acquired using a Bruker Avance 400MHz NMR spectrometer at 300K. The FTIR spectrum was collected using a Nicolet Nexus 870 FTIR spectrometer with a liquid nitrogen cooled MCT detector. The spectrum is an average of 100 scans with 4 cm^{-1} resolution. The polymer film was placed in contact with a Germanium crystal on a "Thunderdome" attenuated total reflectance (ATR) accessory.

Electrochemical Noise Methods (ENM) studies were conducted using a Gamry® PC4 Electrochemical Signal Analyzer (Gamry Instruments, Inc., Willow Grove, PA): ESA 400 Electrochemical Noise System. Eight BAM-PPV coated Al 2024-T3 sample panels were measured. Of these eight sample panels, four panels labeled Q, R, U and V were crossed-scribed and the other four labeled O, N, P and M were non-scribed. Two nominally identical sample panels (i.e., either both scribed or both non-scribed) served as the two working electrodes and a saturated calomel electrode was used as reference. A clamp-on cell was attached to each working electrode (each with an exposed area of 8.5 cm^2) and Dilute Harrison Solution (0.35 wt.% $(NH_4)_2SO_4$, 0.05 wt.% NaCl) was added to each cell (salt-bridge connected the two clamp-on cells). Both the current and voltage noise were measured simultaneously using the ZRA mode. In this mode, the current flow between two identical sample panels was measured and the potential difference between the panels and a reference electrode was also recorded. The data acquisition frequency was 1 Hz and the sampling duration was 2400 s (for a total of 2400 points used to compute the noise resistance).

Polymerization of Monomer

The polymerization of the monomer was performed in a 5L reactor with a mechanical stirrer (Figure 1). The reactor was placed inside a large stainless steel secondary container to allow cooling. Toluene (2L) and THF (2L) was added to the reactor. Dry ice was added to the secondary container with acetonitrile to cool the reactor to -45°C. While the solvent was cooling, 100-150g of monomer was added. When the temperature reached -45°C, 8 molar equivalents of potassium t-butoxide (K-t-OBu) was added (1 mol monomer/8 mol K-t-OBu). The temperature was kept between -45°C and -55°C for 2 hours. The solution was then allowed to warm to room temperature overnight. After 18 hours the reaction mixture was found to be a very viscous orange, semi-gelatinous material. The polymer was precipitated by pouring this material into methanol, filtered through a fine glass frit, and dried under vacuum at 100°C overnight. After drying, the polymer was placed in a soxhlet thimble and extracted with methanol to remove oligomers and some remaining salts, yielding an orange-yellow powder in 93% yield. The extraction was followed by rinsing with methanol, filtering and drying to constant weight. ^{1}H NMR (400 MHz, d_8-toluene): 8.01 (alkenyl), 7.78 (Ar), 3.04 (NCH_2), 2.80 (NCH_3), 1.70 (NCH$_2$C$\underline{H}_2$), 1.37 (N(CH$_2$)$_2$C$\underline{H}_2$), 1.27 (N(CH$_2$)$_3$C$\underline{H}_2$ and N(CH$_2$)$_4$C$\underline{H}_2$), 0.88 (N(CH$_2$)$_5$C$\underline{H}_3$). ^{13}C NMR (100 MHz, d_8-toluene): 148.4 ($\underline{C}_{Ar}$-N), 133.9 ($\underline{C}_{Ar}$-C$_{alkenyl}$), 126.0 ($C_{alkenyl}$), 118.2 (C_{Ar}-H), 57.9 (NCH_2), 42.2 (NCH_3), 32.3 ((N(CH$_2$)$_3$C$\underline{H}_2$), 28.3 (NCH$_2$C$\underline{H}_2$), 27.5 (N(CH$_2$)$_2$C$\underline{H}_2$), 23.2 (N(CH$_2$)$_4$C$\underline{H}_2$, 14.3 (N(CH$_2$)$_5$C$\underline{H}_3$). The NMR and FTIR data were consistent with previously reported results (23). The FTIR spectrum of BAM-PPV is provided in Figure 2 and the peak positions and assignments are shown in Table I. The FTIR spectrum is consistent with the proposed polymer structure.

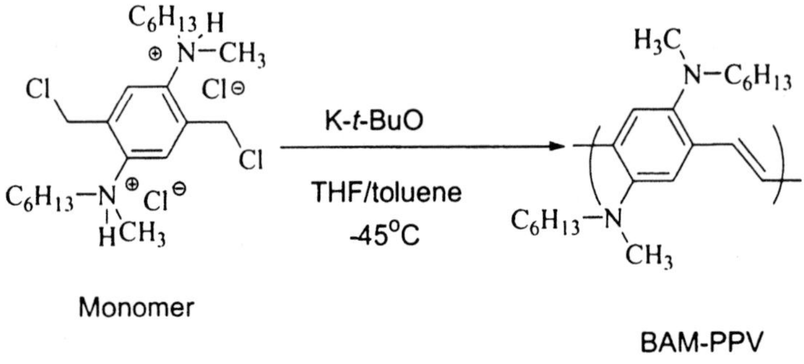

Figure 1. BAM-PPV Polymerization

Results and Discussion

Thermal Properties of BAM-PPV

The polymer was extensively characterized to determine its thermal properties, specifically, glass transition temperature (T_g) and stability in air. Figure 3 shows a low temperature DSC plot of BAM-PPV (2nd heating), which was measured in a nitrogen atmosphere. The glass transition temperature (T_g) for BAM-PPV was measured at 9.0°C.

A small endotherm at this temperature may indicate that the transition has some characteristics of melting, probably due to short side chains becoming "liquid-like." A high temperature DSC plot was run on BAM-PPV (Figure 4). This DSC plot shows that BAM-PPV contains crystal structures that melt near 175°C.

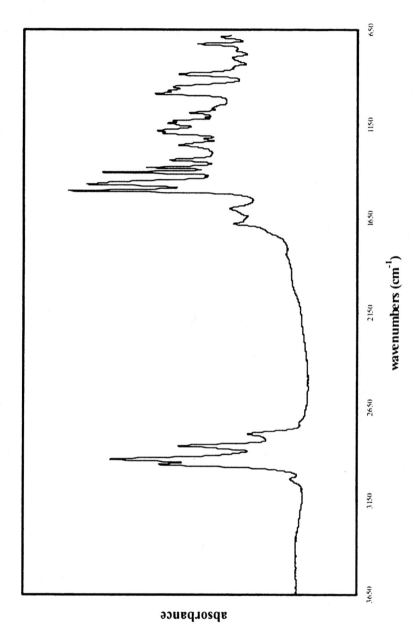

Figure 2. FTIR spectrum of BAM-PPV film

Table I. FTIR Peaks and Assignments for BAM-PPV Polymer Film

Vibrational peaks and proposed peak assignments of BAM-PPV polymer film

peak position (cm^{-1})		assignment	peak position (cm^{-1})		assignment
3032	w	aromatic C-H stretch	1257	m	aromatic C-N stretch
2954	s	asym. methyl C-H stretch	1224	w	
2927	vs	asym. methylene C-H stretch	1192/1179	doublet	C-N aliphatic stretch
2869	sh	sym. methyl C-H stretch	1141	m	
2855	s	sym. methylene C-H stretch	1127	m	
2792	m	C-H stretch (C bonded to N)	1111	w	
1678	w	vinyl C=C stretch	1087	m	
1593	w,b	aromatic C=C stretch	989	s	C-H out of plane on vinyl group
1500	vs	aromatic skeletal vib.	973	sh	
1465	s	asym. methyl def.	953	sh,s	
1455	sh	CH2 scissor	886	s	C-H out of plane on 1,2,4,5 sub. Phenyl
1424	w	N-CH3 sym. methyl bend	797	w	
1403	s		725	m,s	CH$_2$ rock
1377	s	sym. CH3 bend	687	m	
1337	m				

W=weak, m=medium, s=sharp, vs=very sharp, sh=shoulder, b=broad

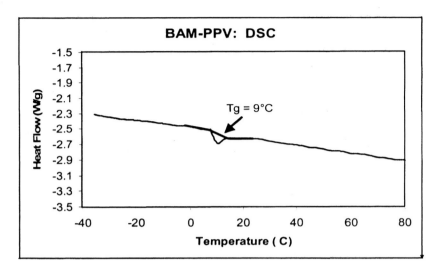

*Figure 3. Low-temperature DSC plot of BAM-PPV
(Reproduced with permission from Reference 24. Copyright 2004
Rapra Technology Ltd.)*

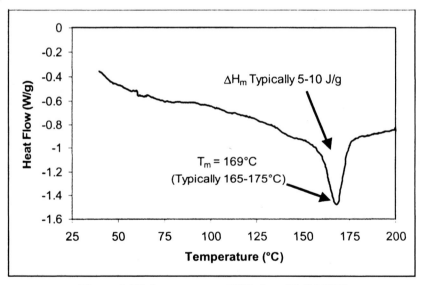

*Figure 4. High temperature DSC plot of BAM-PPV
(Reproduced with permission from Reference 24. Copyright 2004
Rapra Technology Ltd.)*

Thermogravometric analysis, TGA (Figure 5) reveals that initial decomposition begins at 300°C. The 5% weight loss temperature is about 387°C in nitrogen and 320°C in air. This weight loss is associated with a strong exotherm in the DSC traces, which begins at 200-250°C. For this reason, thermal analysis of BAM-PPV is confined to temperatures of 225°C or less in order to ensure the stability of the samples. In a previous publication, a TGA scan of BAM-PPV heated at 120°C for 2 hours, then cooled to room temperature and re-heated to 500°C at 10°C/min, showed complete stability of BAM-PPV at 120°C under nitrogen atmosphere (24, 25). The BAM-PPV polymer is thermally stable up to 225°C without degradation in a nitrogen atmosphere. When BAM-PPV is exposed to air, it will start to degrade about 180°C.

The thermomechanical analysis (TMA) of BAM-PPV using a 0.02 N holding force at temperatures above 10°C, were taken at a heating rate of 10°C/minute (Figure 6). This measurement was to examine the dimensional changes of BAM-PPV. When the BAM-PPV material begins to deform due to softening, there is a lowering of the net expansion rate which introduces error into the signal. Above 50°C, the material is soft enough that compression by the probe exceeds thermal expansion. The TMA of BAM-PPV was repeated using a lower holding force. Although deformation of the sample is reduced, it still takes place. When this is combined with the low holding force, only intermittent contact with the sample takes place above Tg, preventing direct measurement of expansion (Figure 7).

The thermal analysis of BAM-PPV has been carried out to determine the glass transition temperature, thermal decomposition temperature and stability of the polymer. The results have shown BAM-PPV to be thermally stable and can undergo normal processing conditions without degradation, within limits.

Neutral Salt Fog Exposure of BAM-PPV Coated Al 2024-T3

BAM-PPV coated onto Al 2024-T3 coupons was tested as an alternative to CCC in a neutral salt fog chamber. The testing of these substrates followed the ASTM B117 testing procedure for a total of 336 hours (Figures 8 and 9) (26). Any new military pretreatment coating used as an alternative to CCC (Type 1A) must pass 336 hours of neutral salt fog exposure. The coatings must remain intact: no blistering, delamination or evidence of corrosion.

The polymer was dissolved in p-xylene solvent and the solution was then applied via air-brush onto Al 2024-T3 substrates. The BAM-PPV coated Al 2024-T3 panels were dried in a vacuum oven for several hours. After drying the panels were photographed as shown in Figure 8 prior to placement in the neutral salt fog chamber (film thickness ~1.5 μm). The panels were monitored over time and checked for visible signs of corrosion during the neutral salt fog test. A CCC control was also placed in the neutral salt fog chamber and monitored

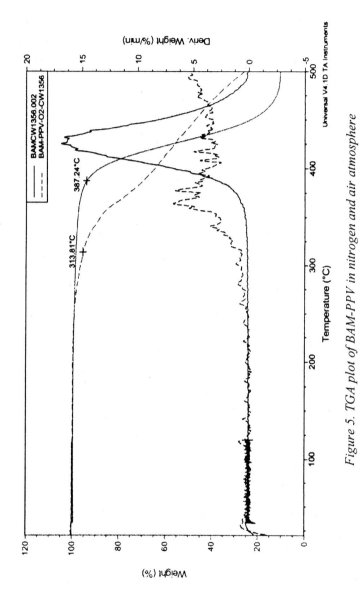

Figure 5. TGA plot of BAM-PPV in nitrogen and air atmosphere

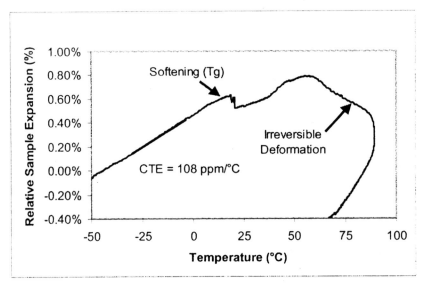

Figure 6. TMA of BAM-PPV at 0. 02N

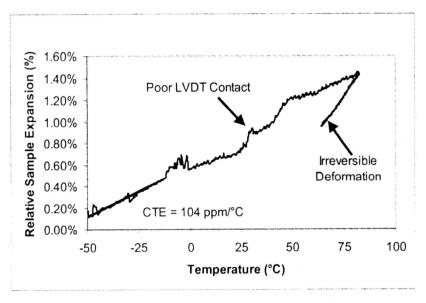

Figure 7. TMA of BAM-PPV at 0.01N

during the test period. At no time did the BAM-PPV coating show delamination, blistering or corrosion. The BAM-PPV coating was able to meet the minimum requirement for passing the 336 hours (Figure 9) neutral salt fog exposure test. The CCC control also showed no signs of corrosion and lasted over 1000 hours in neutral salt fog chamber.

Figure 8. Time = 0 hours   *Figure 9. Time = 336 hours*

(Reproduced with permission from Reference 24. Copyright 2004 Rapra Technology Ltd.) (See page 3 of color inserts.)

Neutral Salt Fog Testing of BAM-PPV Used With Chromium-Free Primers and Topcoats

The BAM-PPV coatings were tested as part of a full military coating system. The polymer solutions were applied onto Al 2024-T3 substrates using an air-brush technique. A standard military epoxy coating, non-chromated primer (epoxy primer for ferrous and nonferrous metals, MIL-P-53022) and urethane-based topcoat (aliphatic urethane, solvent based, MIL-PRF-85285) were applied via spray technique onto the BAM-PPV pretreatment coating on Al 2024-T3. An additional sample set consisted of BAM-PPV as the pretreatment, a chromated epoxy primer (MIL-PRF-23377) and a urethane topcoat (MIL-PRF-85285). A control set consisting of CCC (Mil-C-5541), chromated epoxy primer

(MIL-PRF-23377 Class C) and urethane topcoat (MIL-PRF-85285) were used as controls. All panels were scribed and placed in a neutral salt fog chamber for 2000 hours. They were monitored at selected time intervals (500, 1000, 1500 hours) and at 2000 hours exposure corrosion was evident along the scribed areas for the BAM-PPV military coating system which contained the non-chromated primer (Figure 10). Failure of the panels was observed at 2000 hours. The BAM-PPV pretreatment, chromated epoxy primer and topcoat system showed no sign of corrosion in the scribed area after 2000 hours exposure to neutral salt fog (Figure 11). The control samples were also evaluated to 2000 hours in neutral salt fog chamber without any evidence of corrosion along the scribed area (Figure 12).

Electrochemical Noise Methods (ENM) Studies of BAM-PPV Coated onto Al 2024-T3

ENM was used to evaluate the corrosion protection of the BAM-PPV coated Al 2024-T3 in a long-term immersion study. Noise resistance was measured as a function of immersion time. This experiment was used as a baseline for the evaluation of BAM-PPV containing coating systems; *BAM-PPV is not intended for use without other coating components.* The immersion time was limited to 1008 hours (42 days). The noise resistance of BAM-PPV coated Al 2024 as a function of time is shown Figure 13. The thin BAM-PPV coating (film thickness ~0.4 μm) does not perform as a typical barrier coating (the magnitudes of the measured noise resistances were quite low (10^4 to 10^5 ohm)). However, this performance is consistent with that of other conducting polymer films studied at North Dakota State University and may indicate a surface passivation mechanism.

For the cross-scribed samples (Q&R and U&V), the noise resistance showed an initial value of about 5×10^4 Ω during the first 72 hours of immersion. The resistance dropped to 1×10^4 Ω by 240 hours of immersion and maintained this steady resistance through 1008 hours. At 72 hours, a small amount of corrosion product was observed in the scribe area. By 168 hours, extensive white and flaky corrosion products were observed in and around the defect and also suspended in solution. For the non-scribed sample (P&M and O&N), the average initial noise resistance (1×10^5 Ω) was higher than that of

Figure 10. BAM-PPV + MIL-P-53022 + MIL-PRF-85285
(See page 4 of color inserts.)

Figure 11. BA M-PPV ± MJL-PRF-23377C + MJL-PRF-85285
(See page 4 of color inserts.)

Figure 12. CCC + MIL-PRF-23377C + MIL-PRF-85285
(See page 5 of color inserts.)

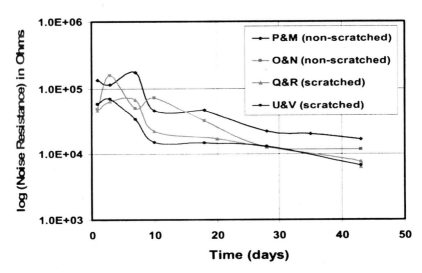

Figure 13: Noise Resistance (Ω, log scale,) of BAM-PPV-coated Al 2024-T3 as
a function of time of immersion (See page 5 of color inserts.)

cross-scribed samples. The resistance increased and reached the maximum at 72 hours (O&N) or 168 hours (P&M), then dropped to about 5×10^4 Ω at 240 hours. The fluctuating noise resistance as a function of immersion time may be due to the formation/breakdown activities of a passive oxide film at the metal/conducting polymer interface. Corrosion products were found on the coating surface after 168 hours of immersion.

The interesting oscillation in noise resistance observed for each of the samples of Figure 13 is similar to that observed for a coating of polyaniline on Al 2024-T3 (27). In that work the values of noise resistance remained between approximately 10^4 and 10^5 Ω cm^2 throughout the immersion period, but an oscillatory behavior in noise resistance was observed during the early stages of immersion. This behavior is markedly different from that of barrier type coatings, which typically show a continually decreasing noise resistance with immersion time. Furthermore, good barrier coating systems typically have high initial noise resistance values, above 10^7 Ω cm^2, which remain high on prolonged immersion. Poor barrier coatings exhibit low initial noise resistance, below 10^5 Ω cm . The oscillations suggest an active interaction between the coating and the metal interface or possibly spontaneous oxidation-reneutralization of the polymer.

Conclusions

The polymerization of monomer to BAM-PPV has been successfully accomplished with high yield. BAM-PPV has been extensively characterized and is shown to be thermally stable up to 225°C without degradation of the polymer in nitrogen and stable up to 180°C in air. The ENM result may indicate that a potential surface passivation mechanism for BAM-PPV pretreatment coatings. BAM-PPV can be applied in a military coating system using standard spray applications. BAM-PPV has been shown to pass the neutral salt fog exposure test for alternative military pretreatment coatings. Although, BAM-PPV used with current nonchromated military coatings showed failure at 2000 hours, when BAM-PPV was incorporated into a military coating containing a chromated epoxy primer and topcoat, the system performed as well as the fully chromated military coating. This represents a step in eliminating the CCC from military coatings and replacing it with a non-chromated pretreatment. Future tests are currently under investigation to optimize these military coatings systems (nonchrome primers) with BAM-PPV pretreatments.

Acknowledgements

The authors would like to acknowledge the continuing support of the Office of Naval Research (ONR), Dr. A. Perez and the Strategic Environmental Research and Development Program (SERDP), Mr. C. Pellerin/Program Manager Pollution Prevention, Dr. J. W. Fischer of NAVAIR and Mr. Craig Matzdorf of NAVAIR-AD for coating of Al 2024-T3 panels.

References

1. Korinek, K.A. In *Chromate Conversion Coatings,* ASM Handbook, Volume 13 Corrosion, Eds. Korb, L. J. and Olson, D. L., ASM International, USA 1998 p. 389.
2. Kendig, M.; Jeanjaquet, S.; Addison, R.; Waldrop, J. *Surface and Coatings Technol.,* **2001**, *140*, 58.
3. Illevbare, G. O.; Scully, J. R.; Yuan, J.; Kelly, R. G.; *Corrosion,* **2000**, *56(3)*, 227.
4. Blasiak, J.; Kowalik, J. *Mutation Research,* **2000**, *469*, 135.
5. National Emissions Standards for Chromium Emissions from Hard and Decorative Chromium Electroplating and Chromium Anodizing Tanks. Environmental Protection Agency. *Federal Register.* RIN 2060-AC14, January 25, 1995.
6. Zarras, P.; Stenger-Smith, J. D. In *Electroactive Polymers for Corrosion Control,* Zarras, P.; Stenger-Smith, J. D.; Wei, Y. Eds.; ACS Symposium Series 843, American Chemical Society, Washington DC; 2003, pp 2-17.
7. Wessling, B.; Posdorfer, J. *J. Electrochemica. Acta.* **1999**, *44*, 2139.
8. Thompson, K. G.; Bryan, C. J.; Benicewicz, B. C.; Wrobleski, D. *Los Alamos National Laboratory Report,* LA-UR-92-360, 1991.
9. Wrobleski, D. A.; Benicewicz, B. C.; Thompson, K. G.; Bryan, B. J. *ACS Polym. Prepr.* **1994**, *35(1)*, 265-266.
10. McAndrew, T.P.; *TRIP,* **1997**, *5(1)*, 7-12.
11. Beard, B.C.; Spellane, P. *Chem. Mater.,* **1997**, *9*, 1949.
12. Fahlman, M.; Jasty, S.; Epstein, A. J.; *Synth. Met.* **1997**, *85*, 1323.
13. Brumbaugh, D. *The AMPTIAC Newsletter,* **1999**, *3(1)*, 1.
14. Yang, S. C.; Brown, R.; Racicot, R.; Lin, Y.; McClarnon, F. In *Electroactive Polymers for Corrosion Control,* Zarras, P.; Stenger-Smith, J. D.; Wei, Y. Eds.; ACS Symposium Series 843 American Chemical Society, Washington DC; 2003, pp. 196-207.
15. Kinlen, P. J.; Ding, Y.; Silverman, D. C. *Corrosion,* **2002**, *58(6)*, 490.
16. Tallman, D. E.; Spinks, G. M.; Dominis, A. J.; Wallace, G.G. *J. Solid State Electrochem.,* **2002**, *6*, 73.
17. Spinks, G. M.; Dominis, A. J.; Wallace, G. G.; Tallman, D. E. *J. Solid State Electrochem.,* **2002**, *6*, 85.

18. Tallman, D. E.; Hie, J.; Gelling, V. J.; Bierwagen, G. P.; Wallace, G. G. In *Electroactive Polymers for Corrosion Control,* Zarras, P.; Stenger-Smith, J. D.; Wei, Y. Eds., ACS Symposium Series 843, American Chemical Society, Washington DC, 2003, pp. 228-253.

19. Stenger-Smith, J. D.; Miles, M. H.; Norris, W. P.; Nelson, J.; Zarras, P.; Fischer, J. W.; Chafin, A. P. U. S. Patent 5,904,990, 1999.

20. Stenger-Smith, J. D.; Zarras, P.; Ostrom, P.; Miles, M. H.; In *Semiconducting Polymers,* Hsieh, B. R.; Wei, Y.; ACS Symposium Series 735, American Chemical Society, Washington DC, 1999, pp. 280-292.

21. Anderson, N.; Stenger-Smith, J. D.; Irvin, D. J.; Guenthner, A.; Zarras, P. Proceedings of the Navy Corrosion Information and Technology Exchange, Louisville, KY, July 16-20, 2001.

22. Anderson, N.; Irvin, D. J.; Webber, C.; Stenger-Smith, J. D.; Zarras, P. *ACS PMSE Prepr.,* **2002,** *86,* 6-7.

23. Stenger-Smith, J. D.; Zarras, P.; Merwin, L. H.; Shaheen, S. E.; Kippelen B.; Peyghambarian, N., *Macromolecules* **1998,** *31,* 7566-7569.

24. Zarras, P.; Anderson, N.; Webber, C.; Guenthner, A.; Prokopuk, N.; Stenger-Smith, J . D.; Polymers in Aggressive and Corrosive Environments, Symposium Proceedings, PACE, September 8-9, 2004, p. 175-181.

25. Zarras, P.; Anderson, N.; Guenthner, A.; Prokopuk, N.; Quintana, R. L.; Hawkins, S. A.; Webber, C.; Baldwin, L.; Stenger-Smith, J. D.; He. J.; Tallman, D. E.; 2005 Tri-Service Corrosion Conference, Orlando, Florida, November 11-14, 2005.

26. ASTM B117, Annual Book of ASTM Standards, ASTM, West Conshohocken, PA, vol 03.02, 2001, p. 1.

27. Tallman, D. E.; Pae, Y.; Bierwagen, G. P. *Corrosion,* **2000,** *56,* 40.

Chapter 11

Oxidation Resistive Coatings at High Temperatures for Iron Cores

Charles A. Sizemore and Chhiu-Tsu Lin[*]

Department of Chemistry and Biochemistry, Northern Illinois University, DeKalb, IL 60115–2862

Several organic-inorganic sol-gel emulsions are formulated and used to produce a thin layer coating (~ 30 nm thick) on iron particles. The coated iron powder and those iron cores molded with the coated iron powder are analyzed by XRD, TG/DTA, and Raman spectroscopy. In TGA graphs, the uncoated elemental iron powder is shown to start a large mass increase at 240-300 °C, whereas those of treated and coated systems have shown to delay to around 500-550 °C. The spectroscopic data indicate that the structure properties of coated and uncoated iron powder are unchanged, suggesting that the electrical properties should remain the same. The relationship between the effective iron powder coatings and their high temperature oxidation resistive properties will be presented and discussed.

153

Introduction

Electrical transformers convert one electrical potential to another electrical potential. In one type of transformer, a series of wire loops is made around an iron core. A second wire is also looped around the core, but with a different number of windings. As current flows through one of the wires, a magnetic field is created inside the iron core. This magnetic field induces a current through the second wire. The change in potential is relative to the proportion of the number of wire loops. For example, if the first wire has 1000 loops and the second wire has 2000 loop, and 10 V is applied to the first wire, then the second wire will acquire a potential of 20 V because there are twice as many loops. These devices also can be used to reduce potential and are used in many appliances to reduce the 110 V potential from the building wiring to a much smaller potential that is more safe (*1*).

As the alternating current cycles, the magnetic field in the iron core also cycles. Eventually, the oscillating magnetic field causes the iron core to increase in temperature. Iron oxidizes spontaneously at room temperature, but this reaction is very slow under most conditions. At increased temperatures, the rate of the corrosion reaction is increased dramatically. Large or frequently used transformer cores are therefore susceptible to oxidation due to high temperatures.

Iron transformer cores are made by applying some heat and large amounts of pressure to iron powder in a mold (*2*). If the individual iron particles in the powder were each coated (a thin, nanometers-thick layer) with a corrosion-resistant, conductive nanocoating, then the iron core produced from the powder should be more corrosion resistant at high temperatures than untreated iron powder (*3*). With this basic idea, the following experiments were performed.

The organic-inorganic hybrid system is currently been developed for oxidative inhibition coatings on metal, ceramic, plastic and paper substrates, (*4-6*) including *In-Situ phosphating coatings* (ISPCs) (*7-9*) and sol-gel chemistry coatings.(*10-12*) In this paper, the nanoscale approach to iron powder coating by sol-gel and ISPCs/sol-gel techniques will be conducted. The oxidation resistance efficiency and chemical and physical properties of the coated iron powder at high temperature will be evaluated. The results are compared to those of the uncoated elemental iron powder and also those of the pretreated commercial iron powder for molding of iron core.

Experiments

Metal and Silicon Alkoxides

Three types of alkoxides were tested for use in the iron particle coatings. The first was a silicon alkoxide, tetramethylorthosilicate (TMOS). The other two were metal alkoxides, titanium (IV) isopropoxide and zirconium (IV) butoxide.

Silanes

Several different silanes were used as the main component of the coating. These included 1,2-bis(triethoxysilyl)ethane (BTSE), 3-glycidoxypropyl-trimethoxysilane (γ-G), 3-aminopropyltrimethoxysilane (γ-A), 3-glycidoxy-propylmethyldimethoxy silane (γ-GD), 3-mercaptopropylmethyldimethoxy silane (D-SH), tetraethylorthosilicate (TEOS), and tetramethylorthosilicate (TMOS). The BTSE, TMOS, and TEOS precursors have four or six potential sites for condensation reactions, which leads to extensive crosslinking and a very hard, three-dimensional coating. The others, γ-A and γ-G, should result in a relatively flexible coating with the silicon end bonded to the metal surface and the epoxy or amino groups oriented outward. These last two were usually used together, with the relative amounts varied to adjust the pH without having to add additional components. These silanes are all acidic in nature, but the γ-A has an amino group at the end, which increases the pH.

When iron is exposed to water and oxygen, corrosion on the microscopic scale begins immediately. This small amount of corrosion usually cannot be seen with the naked eye, but it greatly affects the performance of a coating that is applied on top of it. A metal alkoxide or silicon alkoxide coating simply cannot effectively bond to ionic iron oxide in the same way that it bonds to the non-corroded metal surface. Because the coating is applied in aqueous solution, it is inevitable that some of this "flash rusting" will occur. There are many products and chemicals available commercially that are purported to stop flash rusting, and several of these were tested for compatibility with our system. They include Irgacor 1405 (Ciba), Irgacor 252FC (Ciba), Span 60 Spray (Uniqema), MYRJ 52S Spray (Uniqema), and Wetlink 78 (Silquest).

Application Temperature

The temperature of almost any chemical reaction affects the rate of the reaction. A difference of ten centigrade degrees can be the difference between no observed reaction and a vigorous reaction. In addition, sol-gels are extremely sensitive to temperature and humidity in the manner in which they hydrolyze, coat, and/or cure. For this reason, three different temperatures were used for the coating process of the iron powder.

For the first point, the iron powder was coated at room temperature (21-23°C). For the second and third points, the sol-gel was heated in a hot-water bath until the temperature was 40°C and 60°C, respectively. Higher temperatures than 60°C were not used because the low boiling point of ethanol would change the composition of the sol-gel uncontrollably. Lower temperatures than 20°C were not tested because the coating reaction would, at best, have the same rate as room temperature and would most likely proceed at a slower rate.

Coating Procedure

For this investigation, the sol-gels were prepared in the following way. Metal (or silicon) alkoxides were mixed and sonicated with ethanol. More ethanol was mixed with water to create an aqueous alcohol solution. Any solids used as corrosion inhibitors were dissolved in the aqueous alcohol solution. The metal (or silicon) alkoxide was set to be 5-10 wt% of the total solution (after hydrolysis). The ethanol and water were each used in amounts from 30-60 wt% of the total solution. The corrosion inhibitors were used in very small amounts, usually 0.1-0.5 wt%.

The alkoxide solution and the aqueous alcohol solution were mixed with stirring to begin hydrolysis. For the titanium-based and zirconium-based sol-gels, the hydrolysis reaction was carried out in a cold-water bath. After a few minutes of stirring, the sol-gel was ready to coat the iron powder. For the temperature-varied trials, the sol-gel was heated at this point in a hot-water bath. For all trials, about 300g of sample iron powder (Green Products, Inc., Taiwan) were used for 150g of sol. The powder was added to the sol in a plastic beaker. The solution was stirred for several minutes to coat the iron particles. The powder was removed from the sol-gel by vacuum filtration and was lightly covered with lint-free tissue overnight to dry.

In order to increase bonding between the coating and the substrate, an in-situ phosphatizing reagent, phenylphosphonic acid (PPA),[7-9] was added to some of the sols as 1 wt% of the total solution. The intent was to create phosphorus-oxygen-metal bonds, which are easily formed with heat treatment.

Testing Procedures

The coated and uncoated iron powders were tested in many different ways. Since the primary goal of the research was to produce an iron material that could resist oxidation at higher temperatures, thermogravimetric analysis (Seiko SII SSC/5200 TG/DTA 320) was used to monitor the mass of a sample as the temperature was increased. As the temperature is increased, oxidation of the iron occurs at a more rapid rate. As the iron bonds with oxygen from the air, the sample becomes heavier due to the addition of oxygen atoms; i.e., the metal forms metal oxide. In this manner, a sample can be continually massed as the temperature is slowly increased, and at some temperature, the mass will begin to increase sharply. It was the goal of this work to extend this temperature as much as possible.

For all iron-core coatings, the temperature was increased from room temperature (~20 °C) to 600 °C. Higher temperature testing was desired, but the sample holders (aluminum test pans) melt at temperatures just above 600 °C. The rate of the temperature increase was programmed to be 6 °C/min. This rate was determined to be a compromise between a faster rate (which would decrease the test time from two hours / sample) and a slower rate (which would increase the data resolution).

Other testing methods used for analysis were powder X-ray diffraction (Rigaku Miniflex, Cu K_{α}) and Raman spectroscopy (Renishaw Raman Imaging Microscope System 2000). X-ray diffraction can be used to determine the structure of any crystalline material and was used in this work to analyze the iron powder to see if there were any changes to it after the coating process. Raman spectroscopy is used to elucidate information similar to that obtained from infrared spectroscopy, which gives information about the vibrations of surface coating, for example, sol-gel/metal bonds. It was used to see if there were any differences structurally between any of the coatings. Infrared spectroscopy (IR) (unless a refractive FTIR is used) could not be used in this case because IR depends on the ability of some light to be transmitted through the sample, and these samples would not transmit light because they were mostly metal. Since Raman is a scattering technique, the opaqueness of the sample was not a factor in obtaining good spectra.

X-ray diffraction patterns were obtained for some PPA-treated samples before and after heating to 350 °C for four hours. The samples were heated in a preheated oven on watch glasses. The samples included PPA-treated and untreated samples on both elemental iron powder and commercial iron powder.

Results

Thermogravimetric analysis (TGA) indicates that the performance of the three alkoxides is similar, as shown in Figure 1. Silicon- and zirconium-based coatings prevented large-scale oxidation up to about 470 °C. The titanium-based sol-gel prevented oxidation up to about 480 °C, which may be a negligible difference.

Different silanes used in coating of iron powder gave similar results. As shown in Figure 2, 3-glycidoxypropylmethyldimethoxy silane, 3-mercapto-propylmethyldimethoxy silane, and a mixture of 3-glycidoxypropyltrimethoxy silane and 3-aminopropyltrimethoxy silane all behaved similarly. Also shown is the control, which was simply a sample of the iron powder that was not coated. The γ-GD began oxidation at about 360 °C. The D-SH began to oxidize at about

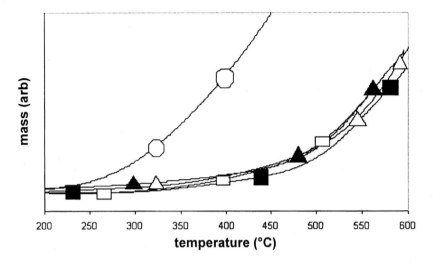

Figure 1: TGA graph of iron particles coated with silicon-based (white triangle), titanium-based (black square), and zirconium-based (white rectangle) sol-gels. Also shown for reference are phosphated iron powder (black triangle) and elemental iron powder (white circle).

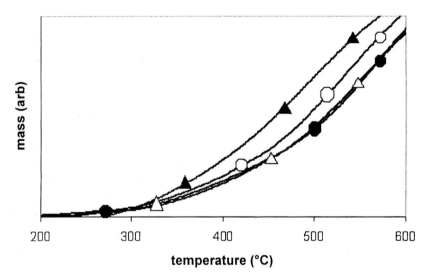

Figure 2. TGA graph of iron particles coated with 3-glycidoxypropyl-methyldimethoxy silane (white circle), 3-mercaptopropylmethyldimethoxy silane (black circle), a mixture of 3-glycidoxypropyltriethoxy silane and 3-amino-propyltrimethoxy silane (white triangle), and the control of the phosphated iron powder sample (black triangle).

375 °C. The mix of γ-G and γ-A oxidized at about the same temperature as the D-SH. All of the silanes resisted oxidation better than the control, which began to oxidize at about 300 °C.

All three alkoxide types were tested at the three different coating temperatures. For the silicon-based coatings, the cooler temperatures were actually better than the 60 °C coating, as shown in Figure 3, although all coatings, again, showed better resistance to oxidation than the uncoated control. The titanium-based coatings shown in Figure 4 do show an improvement at higher temperatures, with the cooler cured coatings showing oxidation at 420 °C and the 60 °C cured coating oxidizing at 450 °C. Unfortunately, the 60 °C sample also shows a loss of mass between 200 °C and 300 °C. Since the loss of mass cannot be attributed to the iron, it must be some degradation of the coating. The results of the TGA of the zirconium-based coatings are given in Figure 5. Like the titanium-based coatings, the zirconium-based coatings seem to improve as curing temperature increases. In addition, like the titanium-based coatings, there is a slight loss of mass between 250 and 300 °C.

Figure 6 shows Raman spectra of coated iron particles and uncoated iron oxide. The coated particles show broader peaks and a new peak at about 665 nm. Figure 7 shows X-ray diffraction results for some coated-iron samples.

Figures 8 and 9 show X-ray diffraction patterns for the PPA-treated samples on both substrates before and after heating at 350 °C against untreated samples. Generally, the samples show a weaker iron pattern after heat treatment, and develop new peaks that indicate the formation of iron oxide. Figure 10 shows the results of the TGA of the PPA-treated samples (B, elemental iron powder, and D, commercial iron powder) and controls (A, elemental iron, and C, commercial iron).

Discussion

Silicon- and zirconium-based sol-gel coatings prevented oxidation of iron particles up to 470 °C. Titanium-based sol-gel coatings may have been slightly better, preventing oxidation up to 480 °C. This result was surprising because titanium (IV) oxide and zirconium (IV) oxide are very thermally stable, with melting points near 2000 °C. For this reason, it was expected that the titanium- and zirconium-based coatings would show a marked improvement over the silicon-based coating. The result indicates that the titanium- and zirconium-based coatings were not better or, at best, marginally better than the silicon-based coating. Because the resistance to oxidation does not appear to be dependent on the material, it could be that the iron particles are not completely coated (or only an ultra thin layer coating at nanoscale). Scanning electron microscopy could be used in the coated samples to confirm or disprove this suggestion.

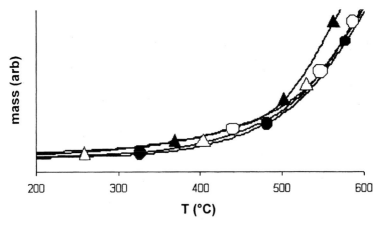

Figure 3. TGA graph of iron particles coated with silicon-based sol-gel and cured at room temperature (black circle), 40 °C (white triangle), and 60 °C (white circle) with phosphated control (black triangle).

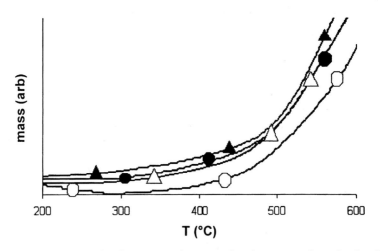

Figure 4. TGA graph of iron particles coated with titanium-based sol-gel and cured at room temperature (black circle), 40 °C (white triangle), and 60 °C (white circle) with phosphated control (black triangle).

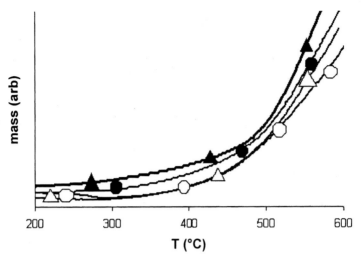

Figure 5. TGA graph of iron particles coated with zirconium-based sol-gel and cured at room temperature (black circle), 40 °C (white triangle), and 60 °C (white circle) with phosphated control (black triangle).

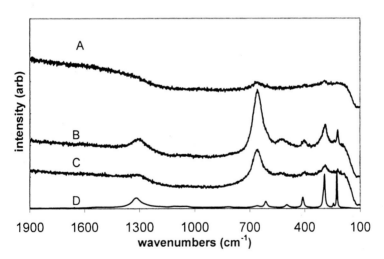

Figure 6. Raman spectra of sol-gel coatings, with silicon-based (A), titanium-based (B), and zirconium-based (C) gels shown with partially oxidized iron sample (D).

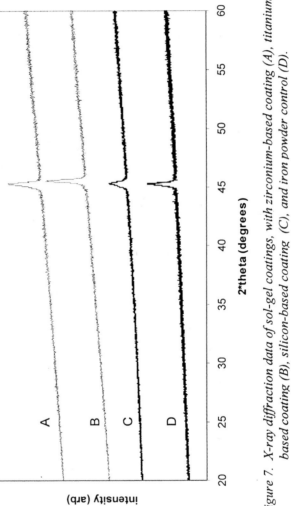

Figure 7. X-ray diffraction data of sol-gel coatings, with zirconium-based coating (A), titanium-based coating (B), silicon-based coating (C), and iron powder control (D).

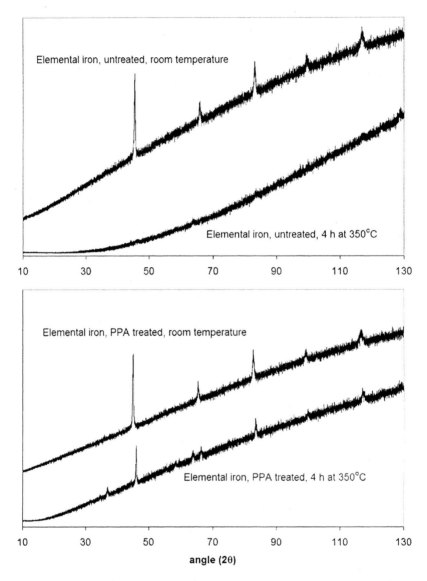

Elemental iron, untreated, room temperature

Elemental iron, untreated, 4 h at 350°C

Elemental iron, PPA treated, room temperature

Elemental iron, PPA treated, 4 h at 350°C

angle (2θ)

Figure 8. X-ray diffraction data of PPA-treated and untreated elemental iron powder.

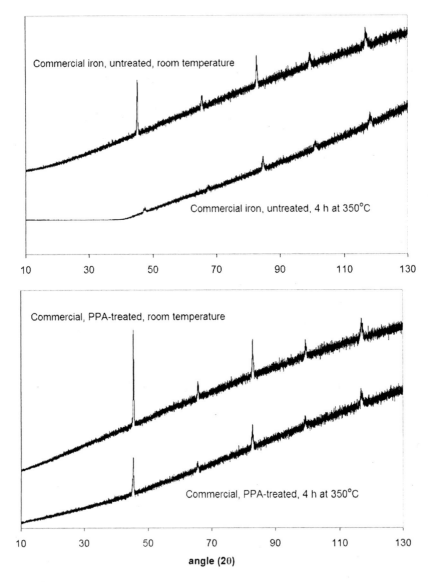

Figure 9. X-ray diffraction data of PPA-treated and untreated commercial iron powder.

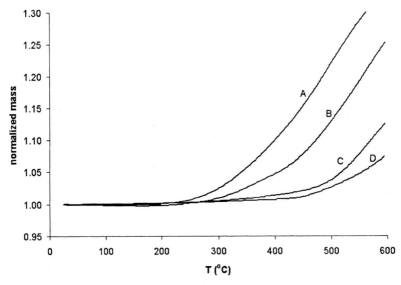

Figure 10. TGA data of (A) untreated elemental iron powder, (B) PPA-treated elemental iron powder, (C) untreated commercial iron powder, and (D) PPA-treated commercial iron powder.

When different silanes were used in the coatings, there was not a significant difference between the tested samples. All began oxidation between 360 and 375 °C, which is lower than the TMOS sample in the first experiment. This result is not as surprising as the result of the first experiment because the organofunctional structures of the tested silanes are not especially thermally stable. TMOS hydrolyzes to form tetrahydroxysilane, which in turn forms silicon dioxide when it is coated on the iron surface. Silicon dioxide, which is essentially glass, can exist at high temperatures (at least 1500 °C) without melting or decomposing. The other tested silanes have organofunctional chains, which are not as thermally stable. These organic parts could be decomposing with heat, which would leave a coating that has no integrity.

The temperature study on the TMOS sol-gel coating showed that the room-temperature coating process was better than the 60 °C coating process. While this result was unexpected, there are a few possible explanations. For one, the increased temperature could have increased the condensation reaction of the sol-gel sols. If this were the case, tiny particles of silicon oligomers would form in solution. Although no particles were observed, they could easily be of the 10-50 nm range, which would eliminate the possibility of seeing them with the human eye. If these particles were formed, they would cause uneven coating of the iron particles, which would expose parts of the iron to oxidation. Another possibility is that the increased temperature caused the acidic solution to attack the metal particles, causing flash rusting. This oxidation could be on a small scale, so that it could not be seen, but still very damaging in that corrosion could take place starting at those sites. The coating would not bond as well at those more ionic sites, complicating the problem. These are only a few suggestions; more work will have to be done to investigate this phenomenon.

The temperature study on the titanium- and zirconium-based sol-gel coatings showed similar effects. The room-temperature process was more effective than phosphate conversion coating only (i.e., the commercial iron powder), but the high-temperature processes were significantly better, especially in the case of the titanium-based coating. This was the expected result, as it was assumed that the binding reaction of the sol-gel to the iron particles was expected to go faster at higher temperature.

The results of the titanium- and zirconium-based sol-gels applied at different coating temperatures are opposite of those of the silicon-based sol-gels with respect to the TGA testing. A possible explanation could be that the increased temperature affects the rates of the reactions of the alkoxides differently. Silicon dioxide, titanium (IV) oxide, and zirconium (IV) oxide are all extremely thermodynamically stable at all the used coating temperatures. However, the relative speeds of the condensation reaction and the metal-binding reaction could be affected differently for different alkoxides. If this is the case, at higher temperatures the titanium- and zirconium-based sol-gels could have an increased rate for the metal-binding reaction while the condensation reaction did not increase to the same degree. The silicon-based sol-gels could have a condensation rate increase that is higher in proportion to the metal-binding reaction.

The Raman and X-ray spectra show that a coating was formed and the metal core remained the same, respectively. This is important for maintaining L and Q values of iron-core devices. Raman is a scattering technique that examines the functional group on the surface of a material. The spectrum is indicative of the structure of the coating and the coating-metal interface. The broad peaks and the presence of a new peak at 665 cm^{-1} indicate that there is an amorphous coating bonded to the metal surface. A product that appeared like the iron / iron oxide mixture (D) would be indicative of iron powder that had been immersed in acidic solution but not coated. The diffraction technique passes X-rays through the iron sample, where they are diffracted and form a pattern based on the crystal structure of the iron. A coating or process that changes the structure of the iron lattice was not desired, and the X-ray data shows that the outer layer of iron remained the same as before it was coated. The single iron peak at 45.3° does not disappear or shift, and no new peaks appear.

The XRD patterns illustrated in Figures 8 and 9 show the degradation of the crystal structure of the iron particles upon heat treatment. At high temperature for an extended period, diffusion of iron atoms causes the crystal to form defects, producing a more amorphous structure. Under these conditions, the diffraction pattern is not as strong, and some iron oxide peaks may begin to appear due to the faster oxidation of the powder. Figure 8 (top) shows the effects of heating on a control of the elemental iron powder. The unheated sample shows typical iron peaks, 2θ at 44.5, 64.8, 82.2, 98.9, and 116.3°. After heating the sample for four hours, the pattern is almost nonexistent, as the sample has lost most of its crystalline character. Figure 9 (top) shows less dramatic but similar results for the untreated commercial iron powder. The peaks in the diffraction pattern of the heated sample are considerably less intense than those in the pattern for the unheated sample. The Figure 8 (bottom) shows the effect of the PPA treatment on the iron powder. Some new peaks are present, indicating the presence of crystalline iron phosphate (or iron oxide), but the iron peaks have not disappeared as they did in the untreated iron powder in Figure 8 (top). Figure 9 (bottom) is analogous in that with heating it shows some decrease in crystalline character, but shows less than that of the untreated commercial iron powder.

The TGA results in Figure 10 show the effect of the PPA treatment on the iron powder's resistance to oxidation. Curve 10a is the untreated elemental iron sample, which has increased its mass by the arbitrary value of 5.0 wt% (corresponding to 13 wt% of the iron oxidized to Fe_3O_4) at 341°C. By contrast, the PPA-treated sample (curve 10b) of the iron does not reach 105 wt% of its initial mass until 408°C. The more practical commercial iron powder (curve 10c) reaches 105 wt% at 519°C, and the PPA-treated version (curve 10d) lasts until 556°C. The results seem to indicate that (i) the PPA additive increases the temperature at which oxidation is rapid, (ii) the difference in oxidation temperature between the commercial iron and elemental iron is larger than the difference between PPA-treated and untreated versions of either substrate, and (iii) either the PPA is less effective on the commercial substrate than the elemental iron ($\Delta T_{com} = 37$°C vs. $\Delta T_{elem} = 67$°C), or the PPA-treatment is less effective at higher temperatures.

Conclusions

Several sol-gel and sol-gel ISPCs coatings were developed that successfully increased the temperature of mass oxidation of iron particles. An arbitrary point of 105 wt% of the starting mass was obtained at 519 °C for the untreated commercial iron powder while the same relative mass was delayed 20-30° for the treated samples. This is a successful achievement because reaction rates can change dramatically with a temperature change of even ten degrees. In this case, the oxidation of the iron powder that occurs even at room temperature was kept to a negligible rate for an additional 20-30°. X-ray diffraction shows that the iron itself did not change structure; therefore, the properties of the iron itself should remain unchanged.

168

Treatments containing an *in-situ* phosphatizing agent (PPA) showed even better results than the other treatments, increasing the 105 wt% point to 37°C higher than the untreated commercial sample. The difference was magnified on pure iron, which gained 67°C by use of the PPA-containing treatment.

References

1. Skoog, D.A., Holler, F.J., Nieman, T.A. *Principles of Instrumental Analysis*; 5th Ed.; Harcourt Brace & Company: Chicago, IL, **1998**.
2. http://www.tscinternational.com/mainferr.html
3. Wang, G., Harrison, A. *J. of Colloid and Interface Science* **1999**, *217*, 203-207.
4. De, .G., Kundi, D. *Chem. Mater.,* **2001**, *13*, 4239-4246.
5. Sanchez, C., Soler-Illia, G., Ribot, F., Grosso, D. *C. R. Chimie,* **2003**, *6*, 1131-1151.
6. Wu, S., Sears, M. T., Soucek, M. D., *Progress in Organic Coatings,* **1999**, *36*, 89-101.
7. Lin, C. T. "Additive package for in-situ phosphatizing paint, paint and method," U.S. patent 5,322,870, **1994**.
8. Neuder, H., Sizemore, C., Kolody, M., Chiang, R., Lin, C. T. *Progress in Organic Coatings,* **2003**, *47*, 225-232.
9. Neuder, H.; Lin, C. T. *J. Coatings Technol.,* **2002**, *74*, 1-5.
10. Van Ooij, W.J., et al. *Surface Engineering,* **2000**, *116*, 386-396.
11. Parkhill, R. L., Knobbe, E. T., Donley, M. S. *Progress in Organic Coatings,* **2001**, *41*, 261-265.
12. Kendig, M. W., Buchheit, R. G. *Corrosion,* **2003**, *59(5)*, 379-400.

Indexes

Author Index

Subject Index